AF306390

Industry 4.0 and the Agricultural Sector in East Asia

Case Studies of Singapore, China and Japan

Industry 4.0 and the Agricultural Sector in East Asia

Case Studies of Singapore, China and Japan

Arthur Pedida
Institute of Technical Education, Singapore

Tai Wei Lim
Soka University, Japan

Yoshihisa Godo
Meiji Gakuin University, Japan

NEW JERSEY • LONDON • SINGAPORE • BEIJING • SHANGHAI • HONG KONG • TAIPEI • CHENNAI • TOKYO

Published by

World Scientific Publishing Co. Pte. Ltd.

5 Toh Tuck Link, Singapore 596224

USA office: 27 Warren Street, Suite 401-402, Hackensack, NJ 07601

UK office: 57 Shelton Street, Covent Garden, London WC2H 9HE

Library of Congress Cataloging-in-Publication Data
Names: Pedida, Arthur author | Lim, Tai-Wei author | Gōdo, Yoshihisa, 1962–   author
Title: Industry 4.0 and the agricultural sector in East Asia : case studies of Singapore,
    China and Japan / Arthur Pedida (Institute of Technical Education, Singapore),
    Tai Wei Lim (Soka University, Japan), Yoshihisa Godo (Meiji Gakuin University, Japan).
Description: [Hackensack, NJ] : World Scientific Publishing Co, [2026] |
    Includes bibliographical references and index.
Identifiers: LCCN 2026013419 | ISBN 9789819830190 hardcover |
    ISBN 9789819830206 ebook | ISBN 9789819830213 ebook other
Subjects: LCSH: Agricultural innovations--Singapore | Agricultural innovations--China |
    Agricultural innovations--Japan | Industry 4.0--Singapore | Industry 4.0--China |
    Industry 4.0--Japan
Classification: LCC S494.5.I5 P44 2026
LC record available at https://lccn.loc.gov/2026013419

British Library Cataloguing-in-Publication Data
A catalogue record for this book is available from the British Library.

For any available supplementary material, please visit
https://www.worldscientific.com/worldscibooks/10.1142/14791#t=suppl

Desk Editors: Murali Appadurai/Pui Yee Lum

Typeset by Stallion Press
Email: enquiries@stallionpress.com

About the Authors

Arthur Pedida is a Senior Lecturer based in Singapore with over a decade of experience in technology education and applied innovation. He was brought up in an agricultural household, where early immersion in farm life influenced his comprehension of food production, resource management, and the lived realities of rural communities. Prior to entering academia, he worked across multiple industries, developing a strong practice-oriented perspective that continues to inform his professional work. His current activities span education, applied research, and innovation, bridging intelligent systems, digital technologies, and sustainable development, with particular relevance to agriculture and food systems. His professional interests center on the responsible application of emerging technologies, systems integration, and the role of education in supporting long-term economic resilience and sustainability.

Tai Wei Lim is a Professor of Business based in Japan with over a decade of multidisciplinary experiences in East Asian studies/history, contemporary China Studies, Sinology, Japanology, and East Asian area studies. He has written on agricultural and resource management topics pertaining especially to Northeast Asia. His immersive and experiential contact with agriculture and other resources is drawn from leading academic and student excursions as well as academic research and observation studies. He teaches sustainable management subjects in Japan and is keen on examining the field of agriculture and resource management through the lens of sustainable management, as well as policy/political science/political economy/international relations (IR), and regionalism perspectives.

Yoshihisa Godo received his PhD degree from the University of Kyoto in 1992. His areas of research include development economics and agricultural economics. Professor Godo's *Development Economics* (3rd edition), co-authored with Yujiro Hayami and published by Oxford University Press in 2005, is especially well known. His book, written in Japanese, *Nihon no Shoku to Nou* (Food and Agriculture in Japan), received the 28th Suntory Book Prize in 2006, one of the most prestigious academic book prizes in Japan. He is also engaged in various social activities, such as the special advisor of Singularity High School. He participates in the International Zheng He Society as an associate member as well as an honorary advisor.

Contents

Introduction

Farmers are the backbone of the food supply chain in East Asia and indeed the world. Enabling the food that they grow improves the welfare and standard of living for both the urban and rural populations of East Asia. As East Asian economies move up the value chain from low-developing to fast-developing economies, from middle-income economies to developed ones, and from fast emerging economies to advanced economies, technologies (especially Industry 4.0 tech) are becoming increasingly important.

Technologies are used not only for increasing quantitative yields but also for sustainable development and mitigating climate change. Agriculture 4.0, based on Industry 4.0, evolved to churn out agricultural crops and food supply by adopting cutting-edge Industry 4.0 technologies, experimenting with emerging technologies in order to have greater efficiency and productivity in the food web, hybridize cross-industry multi-disciplinary technologies, knowhow and applications.

Industrial, garden, urban, rooftop, boutique farming is increasing tap into advancements in Blockchain, Artificial Intelligence, IOT, Big Data technologies, and interoperable cross-platform applications are also becoming more important. Besides production, agriculture is complemented by the food distribution network.

Internet of Things, Big Data Analytics, etc., will be crucial for communication and interoperability between technologies for higher agricultural yields and food security. Problems, challenges, and future research directions will also be discussed in the conclusions of each chapter, as well as the epilogue.

This volume will tap into the practical expertise of Y. Godo and A. Pedida, who are practitioners, observers and researchers in the industry. They are currently engaged in their respective country-specific developments [Japan (Godo) and Singapore/Pedida]. It will also tap into the area studies knowledge of Lim in the Chinese case study.

There will be some comparative analyses between the two case studies based on fieldwork, with other case studies drawn from secondary sources. This is a timely volume, given the crucial issue of recent food supply chain disruptions, growing populations (including the emergence of mega-cities to feed), climate change conditions, an aging population of farmers, and other contemporary challenges. The case studies based on fieldwork value-add to the contents of the volume. The volume also lies at the intersection between Science, Technology, and Society (STS) studies, an interdisciplinary subject area increasingly important as contemporary challenges are increasingly multidiscipline/all-encompassing in nature and require solutions that are not silo-ed into narrower perspectives to tackle them.

Contributions to Existing Literatures

There is no competing volume on this specific topic in the market out there. The current volume contributes to existing literature in the following manner:

- Provides background (cultural/religious/technological contexts) in understanding/interpretation of Industry 4.0 application in the field of agriculture.
- Supplements and complements existing trade professional and non-academic resources. There are a number of journal articles and think tank reports on this subject matter, but not many academic books on the market. For example, one recent 2022 publication, "Trends of the Agricultural Sector in Era 4.0," has a global perspective on the subject matter (Author: Vítor João Pereira Domingues Martinho, https://link. springer.com/book/10.1007/978-3-030-98959-0). It "describes the impacts of Era 4.0 on agriculture around the world, explores the relationships between agriculture 4.0, food 4.0 and industry 4.0, and examines different styles of farming in different world regions".

While this publication has the breadth in coverage, it may not have the same intensity in depth of analysis as the proposed volume. This proposed volume differs in focusing on East Asia and has three in-depth case studies. Out of the three authors, two are practitioners and experts in their areas of study related to this subject matter. And their chapters are based on original fieldwork.

Overall, this is a timely volume, given the crucial issue of recent food supply chain disruptions, growing populations (including the emergence of mega-cities to feed), climate change conditions, an aging population of farmers, and other contemporary challenges.

The next chapter launches into the Singapore Case Study by Art Pedida. It represents the smallest entity in the three selected case studies but the most nimble and most high-tech case study in Southeast Asia, providing a useful comparison with the other two much-larger Northeast Asian comparative examples covered in this volume.

Chapter 1

Singapore Agricultural History

Arthur Pedida

Abstract

As the population continues to rise, urban farmers in Singapore present some challenges to support the country as a self-sufficient nation. Urban farming gained importance as a sustainable solution to a confined metropolis like Singapore, with an increasing public interest in locally sourced food. Local companies are utilizing various farming techniques using vertical farming, hydroponics, and aquaponics, and they've created a multi-dimensional, innovative farming solution. Local farmers, growers, and the rest of the communities are actively involved in cultivating a wide range of crops within the city, contributing to meet their dietary needs, food security, and environmental sustainability.

Introduction

Singapore's agricultural development reflects a dynamic evolution shaped by its strategic geographic position, shifting economic imperatives, and persistent environmental constraints.[1] From an early dependence on trade to the adoption of contemporary urban farming techniques, the nation has

---

[1] Heng, D. (n.d.). Continuities and Changes: Singapore as a Port city Over 700 Years. https://biblioasia.nlb.gov.sg/vol-1/issue1/nov-2005/continuities-changes-port-city/.

consistently adapted its agricultural practices to address the needs of an expanding and transforming population.[2]

In the early 19th century, Singapore's economy was heavily dependent on the cultivation of high-value export crops such as gambier, pepper, and nutmeg.[3] These commodities were central to the island's trade with Europe and China and played a significant role in its early economic development.[4] Gambier, in particular, was highly valued for its application as a dye and tanning agent and emerged as one of Singapore's first major plantation crops.[5] The expansion of the agricultural industry was largely driven by the efforts of local farmers and migrant communities, whose contributions helped lay the foundation of Singapore's farming sector.[6] Conversely, the prioritization of monoculture farming, characterized by extensive cultivation of a single crop, had profound and enduring ecological consequences.[7] This approach led to substantial biodiversity loss, deforestation, and increased soil erosion.[8] As soil fertility deteriorated, agricultural productivity declined, particularly affecting gambier and pepper crops.[9]

Beginning in the 1850s, nutmeg plantations experienced widespread destruction due to a mysterious blight, which blackened branches and killed the fruit, resembling the devastation that had afflicted the original plantings in the 1820s.[10] In response to declining yields and mounting

---

[2] Pwee, T. (2021, April 1). From Gambier to Pepper: Plantation Agriculture in Singapore. BiblioAsia, National Library, Singapore. Retrieved June 11, 2024, from https://biblioasia.nlb.gov.sg/vol-17/issue-1/apr-jun-2021/agriculture/.

[3] *Ibid.*

[4] Hancock, J. and Antan, O. (2022c). The early history of Clove, Nutmeg, & Mace. World History Encyclopedia. https://www.worldhistory.org/article/1849/the-early-history-of-clove-nutmeg--mace/.

[5] Pwee, T. (2021, April 1). *Op.cit.*

[6] *Ibid.*

[7] Balogh, A. (2021, December 13). The rise and fall of monoculture farming. Horizon – the EU Research & Innovation Magazine. Retrieved June 11, 2024, from https://projects.research-and-innovation.ec.europa.eu/en/horizon-magazine/rise-and-fall-monoculture-farming.

[8] DGB GROUP N.V. ("DGB"). (n.d.). DGB | Understanding deforestation: key factors behind forest depletion. https://www.green.earth/deforestation.

[9] *Ibid.*

[10] Pwee, T. (2021, April 1). *Op.cit.*

economic pressures, farmers began to diversify their crops, leading to a surge in pineapple cultivation during the 1920s.[11] This crop thrived in Singapore's climate and was highly profitable for both domestic consumption and export.[12] Rubber, a slower-growing crop, also became a key agricultural product.[13] Farmers capitalized on this by growing pineapples during the long maturation period of rubber trees, ensuring continued income while waiting for their rubber plantations to mature.[14]

Food security emerged as a major concern during the Japanese occupation from 1942 to 1945, as food supplies grew scarce and were tightly rationed.[15] To tackle food insecurity, the "Grow More Food" campaign encouraged residents to grow food in urban areas, converting gardens, roadsides, and vacant plots into productive makeshift farms.[16] This grassroots initiative played a vital role in alleviating food shortages and embodied the community's resilience and ingenuity in navigating the challenges of a period marked by severe scarcity.[17]

Following the war, food security continued to be a challenge across Southeast Asia, including Singapore.[18] The Green Revolution brought technological advancements such as high-yielding rice varieties, which significantly boosted agricultural productivity.[19] The country rapidly expanded its food production, particularly in vegetables and livestock like

---

[11] Gene, N. K. (2022, May 2). Learn about Singapore's history through pineapples, coffee and coconuts. The Straits Times. https://www.straitstimes.com/singapore/community/learn-about-singapores-history-through-pineapples-coffee-and-coconuts.

[12] *Ibid.*

[13] National Library Board Singapore, and Das, A. (n.d.). Pineapple. National Library Board. Retrieved April 12, 2025, from https://www.nlb.gov.sg/main/article-detail?cmsuuid=dffa2769-f0ca-489b-b292-135a3c3a9fb1.

[14] *Ibid.*

[15] Lee, G. B. (n.d.). Wartime Victuals: Surviving the Japanese Occupation. https://biblioasia.nlb.gov.sg/vol-15/issue-1/apr-jun-2019/wartime-victuals/.

[16] Williamson, F. and Goh, J. (n.d.). Growing food in a garden city. Biblio Asia – National Library Singapore. Retrieved April 12, 2025, from https://biblioasia.nlb.gov.sg/nature-and-the-environment/2024/1/singapore-edible-urban-garden/.

[17] *Ibid.*

[18] Low, C.-A. (n.d.). Feeding the Hungry: Children in Post-War Singapore. https://biblioasia.nlb.gov.sg/vol-15/issue-1/apr-jun-2019/feeding-the-hungry/.

[19] *Ibid.*

poultry, pigs, cattle, sheep, and goats, achieving self-sufficiency and exporting surplus poultry to neighboring countries within a few years.[20]

With Singapore's swift rise as a modern city-state, agricultural land gradually gave way to urban infrastructure and high-rise developments.[21] Today, only about 1% of the island's land is dedicated to agriculture.[22] To overcome this challenge, innovative solutions like rooftop farming and agrotechnology parks have been introduced.[23] These parks, equipped with advanced farming techniques such as hydroponics, aquaponics, and vertical farming, maximize productivity in the limited space available and ensure food security in an urban environment.[24]

Food Resilience Through Urban Farming

Singapore's "30 by 30" initiative, launched in 2019, aims to enhance food security by increasing local production to 30% of its nutritional needs by 2030.[25] The Singapore Food Agency (SFA) adopts the strategy of "three food baskets", the foremost being the diversification of food sources.[26] It enhances local food production by funding agricultural research and technology adoption, creatively planning farm spaces, and fostering community support for local farmers.[27]

To facilitate and support the establishment of hi-tech and productive farms in Singapore, Singapore Food Agency (SFA) tenders out land based on qualitative criteria such as production capability, production track record, relevant experience and qualifications, innovation, and

---

[20] Zoe Yeo, "When there were farms" dated 31 January 2019, available at https://biblioasia. nlb.gov.sg/vol-14/issue-4/jan-mar-2019/when-there-wre-farms.

[21] Choo Rui Zhi. (2022, October 12). Of change & challenges: Reminders from Singapore's past agricultural transformations. Singapore Food Agency. https://www.sfa.gov.sg/food-for-thought/article/detail/of-change-challenges-reminders-from-singapore-s-past-agricultural-transformations.

[22] Lim, K. T. (2021, February 10). *Singapore: Food security despite the odds.* Singapore Food Agency. https://www.sfa.gov.sg/food-for-thought/article/detail/singapore-food-security-despite-the-odds.

[23] *Ibid.*

[24] Zoe Yeo, *Op.cit.*

[25] Food. (n.d.-b). Ministry of Sustainability and the Environment. https://www.mse.gov.sg/policies/food.

[26] Food for Thought | Singapore: Food security despite the odds. (2024, June 5). Default. https://www.sfa.gov.sg/food-for-thought/article/detail/singapore-food-security-despite-the-odds.

[27] *Ibid.*

sustainability.[28] This includes vertical farming using hydroponics and aquaponics techniques, as well as innovative urban cultivation setups, such as rooftop farms.[29]

One such example of an innovative approach is Sustenir Agriculture, which has created an indoor vertical farm that can be retrofitted into existing buildings (including office buildings).[30] Sustenir grows vegetables that can't be produced locally, displacing imports and cutting carbon emissions.[31] By leveraging advanced technology, Sustenir can produce up to 3.2 tons of lettuce within a cultivation area of merely 54 square meters, while simultaneously enhancing the taste of its produce, resulting in its signature kale being notably crispier, more flavorful, and sweeter than conventional varieties.[32]

As Singapore advanced its food security goals, it had to integrate climate-responsive and sustainable practices to ensure long-term viability, maintaining suitable humidity and temperatures while developing and implementing new farming methods, and optimizing energy usage and water management.[33] Furthermore, it was essential to integrate sustainable practices that minimized environmental impact, ensuring the long-term viability of the farming systems.[34]

In 2016, during a Parliament Committee of Supply Debate, the Minister of State for National Development and Trade & Industry, Dr. Koh Poh Koon, shared his vision for the future of farming.[35] He highlighted the importance of utilizing advanced technologies to develop

---

[28] *Ibid.*

[29] Diaz, C. (2021, April 7). 3 ways Singapore's urban farms are improving food security. World Economic Forum. Retrieved February 22, 2025, from https://www.weforum.org/stories/2021/04/singapore-urban-farms-food-security-2030/.

[30] *Ibid.*

[31] *Ibid.*

[32] Farm. (2024d, July 4). Default. https://www.sfa.gov.sg/fromSGtoSG/farms/farm/Detail/sustenir-agriculture.

[33] Ministry of Foreign Affairs. (n.d.). Towards a Sustainable and a Resilient Singapore. Sustainable Development Goals. Retrieved April 2, 2025, from https://sustainabledevelopment.un.org/content/documents/19439Singapores_Voluntary_National_Review_Report_v2.pdf.

[34] *Ibid.*

[35] COS 2016: Speech by MOS Dr Koh Poh Koon "Working Together for an Inclusive and Innovative City." (2016, April 11). https://www.mnd.gov.sg/newsroom/parliament-matters/speeches/view/cos-2016-speech-by-mos-dr-koh-poh-koon-working-together-for-an-inclusive-and-innovative-city.

efficient, high-yield, and sustainable agricultural practices.[36] Dr. Koh highlighted that the future of agriculture lies in integrated vertical and indoor farming environments powered by automation and robotics.[37] Dr. Koh added that such farms are expected to achieve high productivity while operating with significantly reduced labour requirements.[38]

This vision is particularly crucial given Singapore's heavy reliance on food imports; the country sources over 90% of its food from around 170 countries and regions, making it vulnerable to supply shocks and global disruptions.[39] In addition to these external risks, climate change further threatens food security by reducing arable land and causing erratic weather patterns.[40] To stay ahead of the situation, Singapore has continuously strengthened its food resilience through diversification and innovation, efforts such as increasing local production, investing in agtech (agricultural technology), and developing alternative food sources align with Dr. Koh's vision for high-tech, self-sustaining farms.[41]

By diversifying its sources of food imports, Singapore reduces its reliance on any single country or region, for example, egg importers now source from Thailand, Poland, and Australia, ensuring a more resilient supply chain.[42] This diversification serves as a buffer against potential disruptions arising from geopolitical tensions, natural disasters, or other unforeseen events.[43]

Expanding into new markets, setting up regional production hubs, and cultivating a globally diverse workforce reduces reliance on any single location, helping to mitigate local disruptions and improve adaptability to evolving global markets and geopolitical shifts.[44] These international agreements help ensure the continuous movement of goods during crises,

---

[36] *Ibid.*

[37] *Ibid.*

[38] *Ibid.*

[39] Singapore's food Challenge. (n.d.-b). https://www.sg101.gov.sg/economy/case-studies/sg-food-challenge/.

[40] *Ibid.*

[41] COS 2016: Speech by MOS Dr Koh Poh Koon, *Op.cit.*

[42] Food for Thought | Staying ready and steady for future food disruptions. (2024, June 5). Singapore Food Agency. https://www.sfa.gov.sg/food-for-thought/article/detail/staying-ready-and-steady-for-future-food-disruptions.

[43] Jeline Chia. (n.d.). SM Teo Chee hean at the Singapore Regional Business Forum 2024. Prime Minister's Office Singapore. https://www.pmo.gov.sg/Newsroom/SM-Teo-Chee-Hean-at-the-Singapore-Regional-Business-Forum-2024.

[44] *Ibid.*

reducing disruptions in the supply chain.[45] Such provisions are vital for maintaining access to essential resources like food in times of emergency.[46] To further strengthen food security, Singapore actively collaborates with global partners by sharing best practices, conducting joint research, and adopting sustainable agricultural technologies.[47]

Singapore has a long history of agricultural research and innovation. A key figure in this development was Henry Nicholas Ridley, who became the first Director of the Singapore Botanic Gardens in 1888. He played a pivotal role in transforming the Gardens into a center for botanical and agricultural advancements.[48] Under his leadership, the Gardens expanded its involvement in botany, agriculture, and forestry. Ridley contributed significantly by publishing the Flora of the Malay Peninsula and establishing an economic garden focused on tropical crops.[49] He was also a pioneering figure in the development of Malaya's rubber industry, conducting research on tapping methods and persuading local planters to cultivate rubber; his efforts led to rubber becoming the region's dominant cash crop.[50] By 1920, the Singapore Botanic Gardens had become a critical source of rubber seeds, contributing to Malaya producing over half of the world's rubber supply. Ridley's innovations not only shaped the region's agricultural landscape but also left a legacy that continues to influence the global rubber industry.[51]

Today, urban farming plays a vital role in the Singapore Food Agency's (SFA) strategy to enhance local food production.[52] Singapore relies on innovative technologies to optimize productivity in a limited

---

[45] Food for Thought | Staying ready and steady for future food disruptions. (2024, June 5). Singapore Food Agency. https://www.sfa.gov.sg/food-for-thought/article/detail/staying-ready-and-steady-for-future-food-disruptions.

[46] Mariam Bint Mohammed Almheiri. (2022, April 17). Food systems need to be more sustainable. Engaging global stakeholders can help. World Economic Forum. Retrieved February 23, 2025, from https://www.weforum.org/stories/2022/08/food-systems-need-to-be-more-sustainable-engaging-global-stakeholders-can-help/.

[47] Singapore's food Challenge. *Op.cit.*

[48] 1888: Ridley and the Malayan Rubber Industry (from 1896). (n.d.). https://www.nparks.gov.sg/sbg/about/our-history/1888-ridley-and-the-malayan-rubber-industry-from-1896.

[49] *Ibid.*

[50] *Ibid.*

[51] *Ibid.*

[52] Food for Thought | Singapore: Food security despite the odds. (2024b, June 5). Default. https://www.sfa.gov.sg/food-for-thought/article/detail/singapore-food-security-despite-the-odds/.

space.[53] To support new entrants, the Starting a Farm: An Industry Guide, launched in 2020, serves as a comprehensive resource that assists individuals and businesses in establishing land-based or indoor farms in Singapore by outlining the necessary steps, regulatory requirements, and best practices to navigate the complexities of farm setup in the city-state.[54] The Singapore Agro-Food Enterprises Federation (SAFEF), a not-for-profit organization established in 2017 to represent and promote Singapore's agro-food sector, which includes livestock, food fish, and vegetable farms, and also collaborates with SFA to simplify and speed up the establishment of urban farms.[55]

One key strategy involves transforming rooftops, underutilized urban spaces, and repurposed industrial areas for farming.[56] Comcrop, Singapore's first-ever commercial rooftop farm, demonstrated this approach.[57] Established in 2011, Comcrop takes urban farming to the next level by integrating vertical farming techniques in city environments.[58] Comcrop, it operates a 500-square-metre rooftop farm at SCAPE, a shopping mall catering to young adults in the heart of Orchard Road's prime shopping district.[59] The company employs a vertical aquaponic system, where waste from live tilapia fish is used to fertilize tomatoes, herbs, and leafy greens growing above.[60]

Innovative urban farming projects continue to expand across Singapore, utilizing spaces such as multi-storey car parks and large-scale rooftop farms. These efforts contribute to Singapore's food security goals

---

[53] *Ibid.*

[54] MSE and SAFEF. (n.d.). Starting a farm: An industry guide. In starting a farm: An industry guide (pp. 4–107). https://www.sfa.gov.sg/docs/default-source/food-farming/industry-guides/sfa-farming-guide_fa-spread-high-resa475fb2b-7755-4c03-aeec-a60f68cc864f.pdf.

[55] *Ibid.*

[56] Food for Thought | Farming in unusual spaces. (2024, June 5). Default. https://www.sfa.gov.sg/food-for-thought/article/detail/farming-in-unusual-spaces.

[57] *Ibid.*

[58] Farm. (2024, July 4). Default. https://www.sfa.gov.sg/fromSGtoSG/farms/farm/Detail/comcrop.

[59] Chandran, R. (2019, January 8). *As Singapore runs out of room, rooftop farms offer solution.* South China Morning Post. https://www.scmp.com/news/asia/southeast-asia/article/2181136/singapore-runs-out-room-rooftop-farms-offer-solution.

[60] *Ibid.*

by making agriculture more efficient and sustainable within a dense urban landscape.[61] One such project, Greenhood Vertical Farm in Hougang, distinguishes itself by cultivating vegetables in a climate-controlled greenhouse.[62] This method eliminates the need for pesticides, as the farm's 1,808 square meters of crops are protected from weather elements.[63]

Greenhood founder and chief executive Gaurav Saraf stepped away from his position as a director at a Japanese bank to dedicate himself to the farm, motivated by his wish to create a tangible and indisputable positive impact on the planet.[64] Saraf noted in an interview with The Straits Times that current agricultural practices face sustainability challenges.[65] Saraf argued that the current food system is inefficient and unsustainable, generating high carbon emissions, excessive food waste, and limiting consumers' access to fresh produce.[66]

The Singapore Food Agency (SFA), as part of its efforts, has awarded $40 million to 12 projects under the second phase of the Singapore Food Story (SFS) R&D Programme, aimed at accelerating food innovation in urban agriculture, aquaculture, future foods, and food safety.[67] Dr. Andy Tay, Presidential Young Professor at the National University of Singapore, is one of the winners under the Seed Grant.[68] His research proposal focused on developing precision genetic tools to produce more vegetables indoors.[69] Dr. Tay explained that his team seeks to strengthen Singapore's food system by applying nanotechnology to enhance plant growth rates and improve resilience to climate-related stresses.[70]

---

[61] Tay, S. (2024, November 13). Under one roof: Vegetable farm and vehicles share a carpark in Hougang. The Straits Times. https://www.straitstimes.com/singapore/under-one-roof-vegetable-farm-and-vehicles-share-a-carpark-in-hougang.

[62] *Ibid.*

[63] *Ibid.*

[64] *Ibid.*

[65] *Ibid.*

[66] *Ibid.*

[67] SFA awards $40 million to accelerate food innovation. (2025, January 3). Default. https://www.sfa.gov.sg/news-publications/newsroom/2024/sfa-awards--40-million-to-accelerate-food-innovation.

[68] *Ibid.*

[69] *Ibid.*

[70] *Ibid.*

Additionally, the Campus for Research Excellence and Technological Enterprise (CREATE), in collaboration with Wageningen University and Nanyang Technological University (NTU), launched a program focused on indoor farming systems.[71] Furthermore, the Singapore Food Agency (SFA) partners with various Institutes of Higher Learning (IHLs) and local farms to develop comprehensive training programs for individuals pursuing careers in the agri-food sector. These initiatives aim to equip students and adult learners with the necessary skills to thrive in the industry.[72]

The Singapore government's introduction of the Agri-Food Cluster Transformation (ACT) Fund, which replaced the Agriculture Productivity Fund (APF), has been instrumental for farms like GroGrace.[73] The APF, particularly through its productivity enhancement scheme, enabled GroGrace to build a proof of concept and leverage funding for capital expenditures, including hardware and software technology.[74]

GroGrace's farm spans 650 sqm, roughly the size of 1½ basketball courts, and boasts four growing floors. It is four times more productive than conventional indoor farms, producing 70 kg of leafy greens per sqm.[75] This transformation was made possible through the support from the Singapore Food Agency's $50 million Agriculture Productivity Fund (APF), as highlighted by Senior Minister of State for Sustainability and the Environment, Ms. Amy Khor, at the ACT Fund launch event.[76]

This support has been crucial in upgrading GroGrace's infrastructure and operational capabilities, helping the farm scale its innovative agricultural practices.[77] The Agriculture Productivity Fund (APF), designed

---

[71] SFA awards $40 million to accelerate food innovation. (2025, January 3). *Op.cit.*

[72] Agri Tech in Singapore: Technologies, jobs & courses [+ FAQs]. (n.d.). Singapore Computer Society. Retrieved February 23, 2025, from https://www.scs.org.sg/articles/agri-tech.

[73] Wee, P. (2024b, March 1). Meet the venture bringing farming indoors in Singapore. https://www.meetings-conventions-asia.com/News/Food-Beverage/Meet-the-venture-bringing-farming-indoors-in-Singapore.

[74] *Ibid.*

[75] Tan, C. (2022, August 3). New urban farm can produce 33 tonnes of leafy greens annually in compact space. The Straits Times. https://www.straitstimes.com/singapore/new-urban-farm-can-produce-33-tonnes-of-leafy-greens-annually-in-compact-space.

[76] *Ibid.*

[77] *Ibid.*

to modernize farming and promote innovation, aligns perfectly with GroGrace's goals of expanding its reach and improving sustainability.[78]

In her speech at the Asia-Pacific Agri-Food Innovation Summit on October 26, 2022, Minister for Sustainability and the Environment, Ms. Grace Fu, reinforced the nation's vision for a sustainable food system.[79] She emphasized the role of innovation, technology, and cross-sector collaboration in ensuring food security amid global challenges like climate change and supply disruptions.[80] A central focus of her address was the "30 by 30" goal and the transformation of Lim Chu Kang into a hub for sustainable food production.[81]

Global food security challenges, such as climate change, are increasingly affecting food production worldwide.[82] Extreme weather conditions, like droughts and floods, stress crops and decrease yields, while excessive rainfall can destroy crops, wash away livestock, and damage food supplies.[83] These disruptions are worsening food insecurity in many regions.[84] In addition, the ongoing war in Ukraine has intensified the global food crisis by disrupting the supply of essential grains and feedstocks.[85]

Ukraine is a major exporter of grain and sunflower products, especially to Europe, Asia, and the Middle East, but the conflict has severely curtailed these exports, causing shortages in Central Asia, Europe, and Africa.[86] These shortages are contributing to rising food prices globally, affecting regions like Asia, where the costs of wheat, corn, and other essential crops are rising sharply.[87]

---

[78] *Ibid.*

[79] Asia-Pacific Agri-Food Innovation Summit – Grace Fu. (n.d.). Ministry of Sustainability and the Environment. https://www.mse.gov.sg/latest-news/speech-by-minister-grace-at-asia-pacific-agri-food-innovation-summit.

[80] *Ibid.*

[81] Food for Thought | A sustainable food system for Singapore and beyond. (2024, June 5). Default. https://www.sfa.gov.sg/food-for-thought/article/detail/a-sustainable-food-system-for-singapore-and-beyond.

[82] *Ibid.*

[83] *Ibid.*

[84] *Ibid.*

[85] *Ibid.*

[86] *Ibid.*

[87] Ukraine's rise in grain and sunflower seed market share limited by ongoing war | Economic Research Service. (n.d.). https://www.ers.usda.gov/amber-waves/2024/may/

Moreover, the energy crisis in Europe has further compounded the agricultural challenges.[88] Higher energy prices have inflated fertilizer costs, making them inaccessible to many farmers, which in turn reduces crop yields.[89] Additionally, transportation costs have soared, further driving up food prices worldwide.[90] Many countries have adopted protectionist measures, such as limiting food exports to ensure domestic supply.[91] These policies can disrupt global markets, leading to inefficiencies and shortages.[92] For instance, Malaysia's export ban on chicken, with its effective date on June 1, 2022, to stabilize domestic supply, has had a direct impact on Singapore's food security, focusing the vulnerability of import-dependent nations in a volatile global food landscape.[93]

Concluding Remarks

Singapore's agricultural trajectory showcased a unique and adaptive response to complex socio-economic transformations, geopolitical pressures, and environmental limitations. Historically rooted in colonial plantation economies centered on gambier, pepper, and nutmeg, the nation's agricultural base was shaped by early export-oriented production that ultimately gave way to diversification as ecological degradation and crop failures became apparent. These shifts marked the beginning of a pragmatic approach to land and resource management that has since become a defining feature of Singapore's agri-food strategy. This adaptability also reflects the broader ethos of Singapore's governance, which prioritizes foresight, strategic planning, and the integration of science and policy to meet evolving national needs.

The evolution of agricultural practices in Singapore demonstrates a broader narrative of resilience and innovation. From the wartime

---

ukraine-s-rise-in-grain-and-sunflower-seed-market-share-limited-by-ongoing-war.

[88] Asia-Pacific Agri-Food Innovation Summit – Grace Fu. *Op.cit.*

[89] *Ibid.*

[90] Energy crisis in Europe hits fertilizer production. (n.d.). https://www.aa.com.tr/en/economy/energy-crisis-in-europe-hits-fertilizer-production/2697912.

[91] Asia-Pacific Agri-Food Innovation Summit – Grace Fu. *Op.cit.*

[92] *Ibid.*

[93] Andres, G. (2022, June 14). Malaysia bans chicken exports: What you need to know. CNA. https://www.channelnewsasia.com/singapore/malaysia-bans-chicken-exports-singapore-supply-price-consumers-2703071.

"Grow More Food" campaign during the Japanese occupation to the technological advances associated with the Green Revolution, the country has continuously redefined its agricultural identity in alignment with prevailing challenges and opportunities. The post-independence era brought about a strategic prioritization of urban development, which required a fundamental change from traditional farming to space-efficient, high-tech agricultural systems. This transition has culminated in the development of agrotechnology parks, vertical farming systems, and aquaculture, all aimed at optimizing productivity within spatial constraints.

Governmental foresight has played a pivotal role in shaping the modern agricultural landscape. Policies such as the "30 by 30" initiative, the establishment of the Agri-Food Cluster Transformation (ACT) Fund, and significant investments in the Singapore Food Story (SFS) R&D Programme display a systemic commitment to enhancing food resilience. These strategies are further supported by infrastructure planning, regulatory facilitation, and public–private partnerships that encourage innovation while ensuring environmental sustainability. Notable examples, including Comcrop, GroGrace, and Sustenir Agriculture, demonstrate the effectiveness of integrated vertical systems, climate-controlled greenhouses, and localized production models in contributing to national food security.

In addition, Singapore's agricultural history offers valuable insights into how a highly urbanized and resource-limited nation can achieve a degree of food sovereignty through innovation, diversification, and institutional support. The nation's capacity to balance economic development with food resilience is particularly instructive considering global disruptions such as climate change, pandemics, and geopolitical conflicts that threaten conventional supply chains. By strategically diversifying import sources, advancing domestic production capabilities, and fostering research collaborations, Singapore not only mitigates its vulnerabilities but also positions itself as a regional and global exemplar of sustainable urban agriculture.

Moreover, the integration of agricultural education, technological research, and public engagement further enhances the robustness of Singapore's agri-food ecosystem. Initiatives such as vocational training, public-private R&D partnerships, and outreach campaigns have contributed to cultivating a new generation of agritech entrepreneurs and informed consumers. The emphasis on environmental stewardship, data-driven farming, and circular economic principles aligns national agricultural development with broader sustainability goals. This convergence of innovation, education, and policy not only addresses immediate concerns of food availability and affordability but also lays the groundwork for a regenerative and inclusive food system in the decades to come.

Lastly, Singapore's agricultural development represents a case study in adaptive governance, technological advancement, and environmental stewardship. The country's transformation from an agrarian trade hub into a smart agri-food innovation leader reinforces the critical role of long-term vision, policy coherence, and cross-sectoral collaboration in addressing food security challenges in the 21st century. As the international community grapples with escalating ecological and geopolitical uncertainties, Singapore's model serves as an instructive framework for reimagining the future of agriculture in highly urbanized, resource-constrained settings.

While agriculture has long been central to societal development and nutritional well-being, it has also contributed substantially to environmental decline.[94] Historical evidence demonstrates that unsustainable agricultural practices can deplete soil fertility, reduce biodiversity, and degrade ecosystems.[95] For example, during the Roman era, monoculture, deforestation, and inadequate land management led to severe soil exhaustion.[96] The resulting nutrient loss compromised agricultural productivity and underscored the need for sustainable practices.[97] Contemporary approaches to sustainability emphasize the preservation of soil health and natural resources to ensure the resilience of food systems.[98] Without such

---

[94] GGI Insights. (2024, October 1). Farming: Cultivating the Foundations of Civilization. Gray Group International. Retrieved March 1, 2025, from https://www.graygroupintl.com/blog/farming.

[95] Kogut, P. (2025, February 28). Soil degradation: Harmful effects & Promising solutions. EOS Data Analytics. https://eos.com/blog/soil-degradation/.

[96] Editor. (2023, April 5). Key factors in the fall of the Roman Empire: unsustainable farming practices and deforestation. DGB Group. Retrieved March 1, 2025, from https://www.green.earth/blog/key-factors-in-the-fall-of-the-roman-empire-unsustainable-farming-practices-and-deforestation.

[97] Soil Erosion and Degradation. (n.d.). World Wildlife Fund. Retrieved March 1, 2025, from https://www.worldwildlife.org/threats/soil-erosion-and-degradation.

[98] Tracextech and Tracextech. (2024, November 6). 7 Best Sustainable Agriculture Practices | Revolutionizing Farming. Blockchain for Food Safety, Traceability and Supplychain Transparency. https://tracextech.com/sustainable-agriculture-practices/.

measures, societies risk facing food insecurity, economic instability, and ecological collapse.[99]

Recent technological advancements are profoundly reshaping the agricultural sector, offering new solutions for enhancing productivity, efficiency, and sustainability.[100] From smallholder farmers to multinational agribusinesses, the integration of emerging technologies is transforming how food is produced, processed, and distributed.[101] The adoption of innovations such as artificial intelligence (AI), blockchain, unmanned aerial vehicles (drones), and the Internet of Things (IoT) has led to significant improvements in crop yields, operational efficiency, and environmental performance.[102] These technologies also strengthen agricultural resilience in the face of climate change and global supply chain disruptions.[103]

Digital technologies further support sustainable agriculture by reducing resource waste, enhancing traceability, and improving food safety.[104] Precision agriculture, robotics, and AI-driven decision-making allow for optimized input management, enabling farmers to allocate water, nutrients, and energy more effectively.[105]

---

[99] History of Agriculture. (n.d.). John Hopkins Center for a Livable Future | Food System Primer. Retrieved June 12, 2024, from https://foodsystemprimer.org/production/history-of-agriculture.

[100] Nijhuis, S. and Herrmann, I. (2019, October 10). The fourth industrial revolution in agriculture. Strategy+Business. https://www.strategy-business.com/article/The-fourth-industrial-revolution-in-agriculture.

[101] Juergen Voegele. (2018, August 3). The Fourth Industrial Revolution is changing how we grow, buy and choose what we eat. World Economic Forum. Retrieved March 1, 2025, from https://www.weforum.org/stories/2018/08/the-fourth-industrial-revolution-is-changing-how-we-grow-buy-and-choose-what-we-eat/.

[102] Nijhuis, S. and Herrmann, I. (2019, October 10). *Op.cit.*

[103] Voegele, J. (n.d.). The Fourth Industrial Revolution is changing how we grow, buy and choose what we eat. World Economic Forum. Retrieved June 12, 2024, from https://www.weforum.org/stories/2018/08/the-fourth-industrial-revolution-is-changing-how-we-grow-buy-and-choose-what-we-eat/.

[104] *Ibid.*

[105] BPM. (2024, October 31). AI in agriculture: Pros, cons and how to stay ahead – BPM. BPM. https://www.bpm.com/insights/ai-in-agriculture/.

Modern agriculture encounters challenges that impact not just individual farms but the entire industry.[106] From small-scale operations to large agricultural enterprises, the global nature of agriculture highlights the importance of collective efforts to address complex issues.[107] In our increasingly interconnected world, building resilient and sustainable agriculture demands collaboration from all sectors of industry.[108]

The World Economic Forum warns that rising global temperatures and escalating climate pressures are likely to intensify water scarcity across the Middle East and North Africa, potentially constraining the region's economic development.[109] In response, there is a growing movement to adopt advanced technologies, such as vertical farming, controlled environment agriculture (CEA) with IoT, and biotechnology, to develop crops that are more resilient to these challenges.[110]

The World Economic Forum's Centre for the Fourth Industrial Revolution, in collaboration with partners in Saudi Arabia and India, is advancing agricultural transformation through the Artificial Intelligence for Agriculture Innovation initiative, which promotes the adoption of AI-driven solutions in farming.[111] This collaboration aims to revolutionize agricultural practices and promote food security and sustainability by partnering with global agtech organizations, exchanging knowledge, and implementing advanced solutions to boost yields and ensure a resilient

---

[106] Sergieieva, K. (2025, February 19). Agriculture problems and technology solutions to them. EOS Data Analytics. https://eos.com/blog/agriculture-problems/.

[107] Goedde, L., Katz, J., Ménard, A., and Revellat, J. (2020, October 9). Agriculture's connected future: How technology can yield new growth. McKinsey & Company. https://www.mckinsey.com/industries/agriculture/our-insights/agricultures-connected-future-how-technology-can-yield-new-growth.

[108] Sustainable Agriculture | BASF Global. (n.d.). https://agriculture.basf.com/global/en/sustainable-agriculture.

[109] Spencer Feingold. (2023, January 26). How can the Middle East and North Africa manage the region's water crisis? World Economic Forum. Retrieved March 1, 2025, from https://www.weforum.org/stories/2023/01/middle-east-north-africa-mena-water-crisis-industry-leaders-solutions/.

[110] Mešić, A., Jurić, M., Donsì, F., Bandić, L. M., and Jurić, S. (2024). Advancing climate resilience: technological innovations in plant-based, alternative and sustainable food production systems. *Discover Sustainability*, 5(1). https://doi.org/10.1007/s43621-024-00581-z.

[111] Agritech: Shaping Agriculture in Emerging Economies, Today and Tomorrow. (2024, April). World Economic Forum. Retrieved June 12, 2024, from https://www3.weforum.org/docs/WEF_Agritech_2024.pdf.

future.[112] Additionally, indoor vertical farming has emerged as a compelling alternative to traditional industrial agriculture, offering significant environmental advantages and improved resource efficiency.[113]

AeroFarms is a pioneer in vertical farming, cultivating crops indoors under control without reliance on soil or weather.[114] Since its founding in 2004 as a Certified B Corporation, AeroFarms has led sustainable agriculture innovation by combining artificial intelligence with plant biology at its flagship facility in Newark, New Jersey.[115] Their proprietary platform ensures high-quality crops with superior flavor, nutrition, traceability, and food safety, driving global advancements in sustainable agriculture.[116] Committed to feeding communities, AeroFarms employs cutting-edge indoor vertical farming techniques and advanced agtech methods to grow crops commercially year-round.[117] Their system incorporates precise controls, including artificial lighting, automated nutrient delivery, and tailored building designs, to create optimal growing conditions.[118]

Indoor farming presents a notable advantage by creating a controlled environment that enables year-round cultivation, unaffected by seasonal changes or unpredictable weather.[119] Bowery Farming, a New York-based company focused on vertical farming and digital agriculture, exemplifies this innovation with its smart farm powered by the advanced Bowery Operating System (BoweryOS).[120] This system precisely tailors growing conditions to meet the specific needs of each plant, ensuring optimal

---

[112] *Ibid.*

[113] Vertical farms: Are they sustainable? (n.d.-b). @RSIS_NTU. https://rsis.edu.sg/rsis-publication/rsis/vertical-farms-are-they-sustainable/.

[114] Joe Myers. (2019, September 4). This company grows crops inside, stacked on top of one another. World Economic Forum. Retrieved March 1, 2025, from https://www.weforum.org/stories/2019/09/vertical-farming-agriculture-aerofarms/.

[115] AeroFarms – Vertical Farming, Elevated Flavor.™. (2025, January 6). AeroFarms the vertical farming, elevated flavor company. AeroFarms. https://www.aerofarms.com/.

[116] Chomsky, R. (2023, October 6). The Future of Agriculture: Exploring AeroFarms vertical farming technology. Sustainable Review. https://sustainablereview.com/the-future-of-agriculture-exploring-aerofarms-vertical-farming-technology/.

[117] *Ibid.*

[118] *Ibid.*

[119] Dupuis, A. (2024b, May 7). Indoor Farming: Sustainable Agriculture's future | Eden Green. Eden Green. https://www.edengreen.com/blog-collection/indoor-farming.

[120] Farming, B. (2023, May 2). Vertical Farming: Why Growing up Can Make a difference. Bowery. https://bowery.co/vertical-farming/.

growth and facilitating seamless scalability.[121] BoweryOS customizes and optimizes crucial factors such as artificial lighting, temperature monitoring, humidity controls, and auto adjustment of nutrient levels tailored to each type of vegetable grown.[122]

This precision-driven approach ensures that every crop receives the ideal conditions necessary for producing high-quality and nutritious yields.[123]

Indoor farming techniques like hydroponics (growing plants in nutrient-rich water), aquaponics (a system combining fish farming and hydroponics), and aeroponics (growing plants with roots suspended in air and misted with nutrients) are great for cities because they don't need soil or large open spaces.[124]

These methods can be applied in vertical farms, on rooftops, or even in indoor spaces, enabling the cultivation of fresh produce close to where people live, reducing transportation costs, ensuring freshness, and promoting local food production in densely populated urban areas.[125]

The modernization of farms has been transformed by technological advancements, including sensors, automation, and information systems that significantly differ from traditional practices of decades past.[126] Sophisticated tools like agricultural robots, GPS technology, temperature and moisture sensors, and aerial imaging now enable farms to operate with greater efficiency, profitability, safety, and environmental sustainability.[127]

Singapore has embraced some of these innovations, with both the government and private sector making substantial investments in

---

[121] Bowery Farming. (n.d.). *Agriculture technology.* https://boweryfarming.com/agriculture-technology.

[122] *Ibid.*

[123] *Ibid.*

[124] Wood, J., Wong, C., and Paturi, S. (2020). Vertical Farming: An assessment of Singapore City. *eTropic Electronic Journal of Studies in the Tropics*, 19(2), 228–248. https://doi.org/10.25120/etropic.19.2.2020.3745.

[125] Godge, M. (2022, September 15). *Hydroponics: Getting to the root of the myths.* Singapore Food Agency. https://www.sfa.gov.sg/food-for-thought/article/detail/hydroponics-getting-to-the-root-of-the-myths.

[126] Agriculture Technology. (2024, December 16). National Institute of Food and Agriculture. Retrieved March 1, 2025, from https://www.nifa.usda.gov/topics/agriculture-technology.

[127] *Ibid.*

cutting-edge solutions like vertical farming.[128] A notable example is Sky Greens, a pioneering Singapore-based company recognized for its advanced vertical farming system.[129] Designed to dramatically boost the production of leafy greens, achieving yields up to five times higher than conventional methods, Sky Greens utilizes tall aluminum frames fitted with rotating planting troughs.[130] This system, developed in collaboration with the Agri-Food and Veterinary Authority (AVA) of Singapore, operates on just 40 watts of electricity (comparable to a single light bulb) to power a 9-meter-tall tower.[131] The water-pulley mechanism rotates the troughs to ensure sunlight exposure and uniform watering, leading to healthier, more consistent growth.[132] Sky Greens' success reflects the impact of research-driven innovation and effective public–private collaboration in advancing sustainable urban agriculture.[133]

Understanding PAR, PPFD and DLI

The study and analysis of Photosynthetically Active Radiation (PAR) are fundamental to optimizing artificial lighting systems for indoor farming, particularly to enhance plant growth.[134] PAR refers to the range of light wavelengths between 400 and 700 nanometers that plants use for photosynthesis.[135] It is commonly measured using Photosynthetic Photon Flux Density (PPFD), which indicates the intensity of light available for plant

---

[128] Food for Thought | A sustainable food system for Singapore and beyond. (2024b, June 5). Default. https://www.sfa.gov.sg/food-for-thought/article/detail/a-sustainable-food-system-for-singapore-and-beyond.

[129] United Nations University. (n.d.). Farming in the sky in Singapore. https://ourworld.unu.edu/en/farming-in-the-sky-in-singapore.

[130] Wood, J., Wong, C., and Paturi, S. (2020c). Vertical Farming: An assessment of Singapore City. *eTropic Electronic Journal of Studies in the Tropics*, 19(2), 228–248. https://doi.org/10.25120/etropic.19.2.2020.3745.

[131] *Ibid.*

[132] Begum, S. (2019, June 11). Vertical farm receives the world's first urban farm certification for organic vegetables. The Straits Times. https://www.straitstimes.com/singapore/vertical-farm-receives-the-worlds-first-urban-farm-certification-for-organic-vegetables.

[133] Farming in the sky in Singapore – our world. *Op.cit.*

[134] Team, Z. (2022, November 18). PAR, LEDs, and indoor farming. ZipGrow Inc. https://zipgrow.com/photosynthetically-active-radiation-par-and-indoor-farming/.

[135] Grow lights. (n.d.). https://gardeningsg.nparks.gov.sg/page-index/horticulture-techniques/grow-lights/.

growth.[136] For light-demanding edible crops, specialized LED grow lights with higher PPFD values are especially effective.[137]

Within the PAR spectrum, blue light (400–500 nm) plays a crucial role in regulating plant growth and development.[138] It is essential for photosynthesis and strongly influences plant structure by inhibiting stem elongation, resulting in shorter, more compact plants.[139] Blue light also governs photomorphogenesis, affecting processes like stomatal opening, chloroplast movement, and flowering timing.[140] In general, blue light leads to plants with thicker, smaller, and darker green leaves.[141]

Red light, with wavelengths between 620 and 700 nm, particularly at 630 and 660 nm, is crucial for plant growth, as it is efficiently absorbed by chlorophyll to promote flowering, as well as stem and leaf development.[142] In contrast, far-red light suppresses germination by signaling low sunlight, promotes excessive stem elongation and leaf expansion during vegetative growth, but when strategically applied at the end of the light cycle, it can accelerate flowering and significantly increase annual yields.[143]

At LightSym 2024, Dr. Bruce Bugbee examined the historical progression of horticultural lighting technologies and emphasized the critical role of far-red light (700–750 nm) in promoting plant

---

[136] *Ibid.*

[137] *Ibid.*

[138] Runkle, E. (2017). Effects of blue light on plants. In GPNMAG.COM. https://www. canr.msu.edu/floriculture/uploads/files/blue-light.pdf.

[139] Marco Landi, Marek Zivcak, Oksana Sytar, Marian Brestic, Suleyman I. Allakhverdiev, Plasticity of photosynthetic processes and the accumulation of secondary metabolites in plants in response to monochromatic light environments: A review. *Biochimica et Biophysica Acta (BBA) – Bioenergetics*, Volume 1861, Issue 2, 2020, 148131, ISSN 0005-2728, https://doi.org/10.1016/j.bbabio.2019.148131.

[140] *Ibid.*

[141] Effects of Blue Light on Plants. (n.d.). Michigan State University – College of Agriculture & Natural Resources. Retrieved June 11, 2024, from https://www.canr.msu. edu/floriculture/uploads/files/blue-light.pdf.

[142] MarsHydro. (2023, May 20). What Grow Light Features Contribute the Most To The Flowering Stage. Mars Hydro. Retrieved March 1, 2025, from https://www.mars-hydro. com/info/post/two-features-of-grow-lights-that-contribute-the-most-to-the-flowering-stage.

[143] Lightworks, C. (2025, April 22). How does the far red light spectrum affect plants? California LightWorks. https://californialightworks.com/blog/how-does-the-far-red-light-spectrum-affect-plants/.

development, advocating for its formal inclusion within the definition of photosynthetically active radiation (PAR) based on empirical findings using Apogee Instruments' precision measurement tools.[144] According to Dr. Bruce Bugbee, far-red light is essential for enhancing photosynthetic efficiency, promoting plant elongation and morphological development, and accelerating flowering, thereby serving as a critical factor in optimizing plant growth within controlled environment agriculture.[145]

Overall, photosynthetically active radiation (PAR) significantly influences plant photosynthesis, growth, development, morphology, and metabolism, with varying effects depending on light wavelengths, plant variety, growth stage, and its relevance to studies on plant growth, photosynthesis, and vegetation indices.[146]

A reliable way to determine how much photosynthetically active radiation (PAR) reaches a plant's surface over time is by measuring its Photosynthetic Photon Flux Density (PPFD).[147] PPFD, expressed in micromoles per square meter per second (μmol/m²/s), indicates the number of photosynthetically active photons available for photosynthesis, essentially showing how much usable light surrounds the plant.[148] This metric is crucial because it directly reflects the amount of light a plant can absorb for growth. Higher PPFD values mean more usable light is reaching the plant.[149] PPFD values are often listed in grow light specifications.[150]

---

[144] Dr. Bugbee's Expert Insights at LightSym. (n.d.). Apogee Instruments, Inc. https://www.apogeeinstruments.com/apogee-instruments-blog/dr-bugbees-expert-insights-at-lightsym/.

[145] *Ibid.*

[146] Photosynthetically Active Radiation (PAR) – Renke. (n.d.). Environment Monitoring Sensors Manufacturer. https://www.renkeer.com/photosynthetically-active-radiation.

[147] Measurement of PAR (Photosynthetically Active Radiation) GigaHertz-Optik. (n.d.). https://www.gigahertz-optik.com/en-us/service-and-support/knowledge-base/measurement-of-par/.

[148] Marketing. (2025a, January 16). Untangling the differences between PAR, PPF, PPFD, and PFD | UPRTEK. UPRtek. https://www.uprtek.com/en/blogs/untangling-par-ppf-ppfd-pfd.

[149] *Ibid.*

[150] Monnit Corporation. (2023, October 5). *How PPFD measurement fuels crop light monitoring.* https://support.monnit.com/article/83-what-are-ppfd-and-ppf-measurements.

To measure PPFD accurately in different environments, outdoors, in greenhouses, or indoors, quantum sensors are used.[151] These sensors are ideal for continuously monitoring the light plants need for photosynthesis.[152] Understanding PPFD distribution aids farmers in choosing the most effective lighting systems to support their plants' photosynthetic requirements.[153]

The Daily Light Integral (DLI) is a key metric used to measure the total amount of photosynthetically active radiation (PAR) that a plant receives over the course of a day, expressed in moles per square meter per day (mol m^{-2} d^{-1}).[154] DLI is a crucial metric to measure in every greenhouse as it directly impacts plant growth, development, yield, and quality.[155] It accounts for both the intensity of light and the duration of exposure, offering a comprehensive understanding of a plant's daily light intake.[156] Monitoring DLI is essential in greenhouse environments, as it directly influences plant growth, development, yield, and overall quality.[157] To determine DLI, quantum meters are used to measure the photosynthetic photon flux density (PPFD), which is then averaged over time to calculate the total light exposure across a 24-hour period.[158]

The Daily Light Integral (DLI) is a valuable metric that measures the total amount of light a plant receives over a 24-hour period, capturing both light intensity and duration.[159] It highlights the importance of light

---

[151] Apogee Instruments, Inc. (n.d.). *Applications and uses of quantum sensors.* https://www.apogeeinstruments.com/applications-and-uses-of-quantum-sensors/.

[152] *Ibid.*

[153] Kris. (2019, January 24). Horticulture Light for Growers – PAR, PPF, and PPFD explained. Dimlux Lighting – the Best Grow Lights. https://www.dimluxlighting.com/knowledge/blog/horticulture-light-terms-explained.

[154] Ariana P. Torres, Roberto G. Lopez, Commercial Greenhouse Production – Measuring Daily Light Integral in a Greenhouse, Purdue University. https://www.extension.purdue.edu/extmedia/ho/ho-238-w.pdf.

[155] *Ibid.*

[156] Runkle, E. (2006). Daily Light integral defined. GPN. https://www.canr.msu.edu/uploads/resources/pdfs/dailylightintegraldefined.pdf.

[157] Daily Light Integral: Measuring Light for Plants. (n.d.). Apogee Instruments, Inc. https://www.apogeeinstruments.com/daily-light-integral-measuring-light-for-plants/.

[158] *Ibid.*

[159] Torres, A. P. and Lopez, R. G., Purdue University, Purdue Department of Horticulture and Landscape Architecture, Purdue Floriculture, Purdue extension, Faust, J., Clemson University, Mapping monthly distribution of daily light integrals across the contiguous

quantity and quality in driving photosynthesis and supports optimal plant growth and development.[160] By analyzing the intensity and timing of light exposure, horticulturists and farmers gain a deeper understanding of its overall impact on plant physiology.[161] This insight enables the creation of customized cultivation strategies that enhance yield and improve photosynthetic efficiency across various agricultural systems.[162] As a result, DLI has become a vital tool in modern, precision-driven farming that relies on a nuanced understanding of plant-light interactions.[163]

In an indoor farming environment, optimizing plant growth depends on strategic light placement that ensures uniform distribution, appropriate intensity, balanced spectral composition, and controlled photoperiods to promote consistent development and maximize yields.[164] In the article "Stacked 'Vertical' Farming Doesn't Stack Up," Dr. Nate Storey explains that Plenty's "true vertical" farming design allows for the use of high-intensity lighting, enhancing photosynthesis and crop yields, by facilitating efficient heat dissipation through vertical airflow, unlike traditional stacked tray systems that trap heat and limit light intensity.[165] Dr. Nate Storey explained that expanding the cultivation surface enhances consistency in plant development and simplifies maintenance, as more uniform humidity and temperature conditions at the canopy help promote healthier, low-stress growth.[166]

United States, & Pamela C. Korczynski, Joanne Logan, and James E. Faust. (2002). Measuring daily light integral in a greenhouse. In Purdue extension (pp. 2–3). https://www.extension.purdue.edu/extmedia/ho/ho-238-w.pdf.

[160] Daily Light Integral: Measuring Light for Plants. *Op.cit.*

[161] Greer, D. H., Robinson, S. A., Watling, J. R., Bittisnich, D. J., Fukai, S., Beadle, C. L., and Kriedemann, P. E. (2017). Chapter 12-Sunlight and plant production. https://www.semanticscholar.org/paper/Chapter-12-Sunlight-and-plant-production-Greer-Robinson/25b16d14741d7fbdd3f1c4774257c004d25e0185.

[162] Daily Light Integral: Measuring Light for Plants. *Op.cit.*

[163] Daily Light Integral: Measuring Light for Plants. *Op.cit.*

[164] Vejgaard, N. (2022, June 2). The right lights for vertical farming types and techniques – LED iBond. *LED iBond.* https://ledibond.com/the-right-lights-for-vertical-farming-types-and-techniques/.

[165] *Stacked "vertical" farming doesn't stack up.* (2023, December 1). *Vertical Farm Daily.* https://www.verticalfarmdaily.com/article/9582342/stacked-vertical-farming-doesn-t-stack-up/.

[166] NASA Research launches a new generation of indoor farming | NASA spinoff. (n.d.). https://spinoff.nasa.gov/indoor-farming.

Fourth Industrial Revolution (4IR)

The Fourth Industrial Revolution (4IR) is transforming agriculture by integrating cutting-edge technologies such as the Internet of Things (IoT), artificial intelligence (AI), big data analytics, robotics, and cloud computing.[167] Precision agriculture, central to this transformation, relies on sensors, drones, and GPS to gather real-time data on soil conditions, crop health, and weather patterns.[168] This data-driven approach enables farmers to optimize resource use, applying water, fertilizers, and pesticides precisely when and where needed, reducing waste and environmental impact while maximizing yield.[169] These technologies also enable smarter decision-making at every stage of the farming cycle, leading to increased efficiency and reduced labor costs.[170]

Beyond resource efficiency, 4IR technologies enable smarter, faster decision-making throughout the farming cycle, increasing productivity and reducing labor costs.[171] Meanwhile, innovations like blockchain are enhancing food supply chain transparency and traceability, allowing consumers to verify product origins and quality, building trust and potentially raising returns for producers.[172] E-commerce platforms also allow farmers

---

[167] Klaus Schwab. (2016, January 14). The Fourth Industrial Revolution: what it means, how to respond. World Economic Forum. Retrieved March 2, 2025, from https://www.weforum.org/stories/2016/01/the-fourth-industrial-revolution-what-it-means-and-how-to-respond/.

[168] Guebsi, R., Mami, S., and Chokmani, K. (2024). Drones in precision agriculture: A comprehensive review of applications, technologies, and challenges. *Drones*, 8(11), 686. https://doi.org/10.3390/drones8110686.

[169] path-intraafrica.org. (2024, June 12). The benefits of Precision Agriculture for sustainable farming. https://path-intraafrica.org/the-benefits-of-precision-agriculture-for-sustainable-farming/.

[170] Gammanpila, H. W., Sashika, M. a. N., and Priyadarshani, S. V. G. N. (2024). Advancing horticultural crop loss reduction through robotic and AI technologies: Innovations, applications, and practical implications. *Advances in Agriculture*, 2024(1). https://doi.org/10.1155/2024/2472111.

[171] Abdo Hassoun, Iman Dankar, Zuhaib Bhat, Yamine Bouzembrak, Unveiling the relationship between food unit operations and food industry 4.0: A short review, *Heliyon*, Volume 10, Issue 20, 2024, e39388, ISSN 2405-8440, https://doi.org/10.1016/j.heliyon.2024.e39388.

[172] Duong, C. D. (2025). Blockchain technology in the organic food supply chain: Insights from an integrated model of trust transfer theory and theory of planned behavior. *Journal of Global Marketing*, 1–21. https://doi.org/10.1080/08911762.2025.2477007.

to access wider markets directly, reducing dependence on intermediaries and increasing income opportunities.[173]

Agriculture 4.0, a term used to describe this tech-driven evolution, digitizes agriculture by integrating AI, IoT, robotics, and big data to improve sustainability, resilience, and efficiency.[174] More than just adopting technology, the development seeks to build a smarter, more inclusive agricultural system that empowers farmers, enhances food security, and minimizes the sector's environmental footprint.[175]

However, realizing this vision globally requires overcoming significant barriers, limited digital infrastructure, gaps in digital literacy, and the need for inclusive access to tools and training, all of which present major challenges.[176] Bridging these divides is essential to ensure that the benefits of Agriculture 4.0 are shared equitably across all regions and farm sizes.[177]

Since 2015, hunger and food insecurity have been rising at an alarming rate, driven by the combined effects of the COVID-19 pandemic, armed conflicts, climate change, and widening inequalities.[178] A detailed report from the United Nations Food and Agriculture Organization (FAO) warns that these issues are likely to worsen, especially with the global population projected to increase by 2 billion by 2050.[179]

---

[173] Market.Us. (2024, November 5). E-Commerce of Agricultural Products Market—A Digital Green Revolution! Global Trade Magazine. https://www.globaltrademag.com/e-commerce-of-agricultural-products-market-a-digital-green-revolution/.

[174] Agricultural 4.0 Leveraging on Technological Solutions: Study for smart farming sector. (n.d.). https://arxiv.org/html/2401.00814v1.

[175] *Ibid.*

[176] Agriculture 4.0: Empowering Farmers with Data. (2023, November 2). Agri Business Review. https://www.agribusinessreview.com/news/agriculture-40-empowering-farmers-with-data--nwid-1068.html.

[177] Sara Pantuliano. (2020, January 20). Four ways governments can leverage 4IR to achieve the SDGs. World Economic Forum. Retrieved March 2, 2025, from https://www.weforum.org/stories/2020/01/governments-leverage-4ir-achieve-sdgs/.

[178] Martin. (2023, October 19). Goal 2: Zero hunger – United Nations Sustainable Development. United Nations Sustainable Development. https://www.un.org/sustainabledevelopment/hunger/.

[179] Nuccitelli, D. and Nuccitelli, D. (2022, October 20). UN report: The world's farms stretched to "a breaking point." Yale Climate Connections. https://yaleclimateconnections.org/2022/01/un-report-the-worlds-farms-stretched-to-a-breaking-point/.

Compounding this crisis is the degradation of farmland caused by desertification and other human and climate-driven impacts, which have significantly reduced the amount of arable land available for cultivation, posing a serious threat to global food security.[180] In this context, stopping deforestation and possibly returning land to nature is a more sensible long-term environmental responsibility.[181]

Advancements in seed breeding and biological stimulants (akin to plant "vitamins") now help crops better endure extreme weather such as droughts, heatwaves, and floods.[182] These technologies also improve shelf life and reduce food waste by enhancing the durability and appeal of fruits and vegetables.[183]

Despite these breakthroughs, adoption remains limited; a 2022 McKinsey Global Farmer Insights Survey found that fewer than 5% of farmers in Asia, Europe, and the Americas use next-generation technologies, while 21% use basic farm management software.[184] One example of advanced adoption is automated precision spraying, where sensors and real-time field data enable equipment to adjust chemical volume and timing based on crop gaps.[185]

Emerging systems can even use real-time weed recognition via computer vision: a camera captures field images, AI identifies weeds, and nozzles deliver targeted micro-doses of herbicide, reducing chemical use significantly.[186] Drones also support aerial crop monitoring, offering farmers detailed visualizations of crop health, pest infestations, and growth patterns.[187] Predictive analytics further improve outcomes by forecasting potential pest or disease outbreaks, enabling early, targeted

---

[180] *Ibid.*

[181] Fyrwald, J. E. (2022, January 3). How farming innovations can feed the world and protect the planet. World Economic Forum. Retrieved April 10, 2025, from https://www.weforum.org/stories/2022/01/how-farming-innovations-feed-world-protect-planet/.

[182] *Ibid.*

[183] *Ibid.*

[184] Bland, R., Ganesan, V., Hong, E., and Kalanik, J. (2023, May 31). Trends driving automation on the farm. McKinsey & Company. https://www.mckinsey.com/industries/agriculture/our-insights/trends-driving-automation-on-the-farm.

[185] *Ibid.*

[186] Piddubna, A. (2024b, July 9). Intelligent spraying in agriculture for precise herbicide application. Intellias. https://intellias.com/smart-spraying-technology-in-agriculture-for-precise-herbicide-application.

[187] The Farming Insider. (2025, February 3). The future of farming trends in automated machinery. https://thefarminginsider.com/future-farming-automated-machinery/.

responses and improving overall productivity; these technologies are no longer theoretical, they are already in use.[188]

In Singapore, Kok Fah Technology Farm (KFTF) demonstrates the power of automation with equipment such as Automated Seeding Machines and Bale Dosage Bunkers.[189] These tools streamline operations, reduce labor needs, and support sustainability through systems like composting, which converts vegetable waste into high-grade compost.[190] Kok Fah Technology Farm's Greenhouse Nursery further optimizes growing conditions for young plants.[191]

The future of agtech lies in upskilling, as the industry requires a skilled workforce to support high-value jobs, and mid-career workers eager to learn new skills may be key to driving growth.[192] A smart farmer leverages technologies such as sensors for soil, water, and temperature management, specialized software for farm-specific needs, advanced connectivity like cellular and LoRa, GPS and satellite for precise location tracking, robotics like autonomous tractors, and data analytics solutions to optimize farm operations and drive productivity.[193]

On January 14, 2022, Workforce Singapore (WSG), the Singapore Food Agency (SFA), and Republic Polytechnic (RP) launched the Career Conversion Programme (CCP) for the Agri-tech sector.[194] The program aims to benefit 100 individuals over two years, fostering a local talent pipeline that supports the Singapore Green Plan and the "30 by 30" goal of producing 30% of Singapore's nutritional needs locally by 2030.[195]

---

[188] *Ibid.*

[189] Best Vegetable Farm with Weekend Wholesale Market in Singapore – 2024 – Kok Fah Technology Farm Pte Ltd, https://kokfahfarm.com.sg/.

[190] *Ibid.*

[191] Singapore Agro-Food Enterprises Federation Limited. (2020, November 30). *Contributing to Singapore's "30 by 30" goal – Dave Huang of Kok Fah Technology Farm.* Retrieved March 2, 2025, from https://safef.org.sg/blog/contributing-tosingapores-30-by-30-goal-dave-huang-of-kok-fah-technology-farm/.

[192] Republic Polytechnic. (2026). *Agriculture research and innovation (AGRI) centre.* https://www.rp.edu.sg/industry/solutions/centres/agri/.

[193] What is Smart Farming? It's The Future of Agriculture | IoT For All – 2024 – SciForce, https://www.iotforall.com/smart-farming-future-of-agriculture.

[194] Reimagining the Agri-Tech workforce. *Op.cit.*

[195] New career conversion programme for agri-tech sector | WSG. (n.d.). Workforce Singapore. https://www.wsg.gov.sg/home/media-room/media-releases-speeches/new-career-conversion-programme-for-the-agri-tech-sector-to-benefit-100-mid-career-individuals-by-2023.

Singapore's high-tech agriculture sector is expected to create 4,700 jobs by 2030, with 70% of these roles targeting professional, managers, executives, and technicians (PMETs) or skilled laborers.[196] Under the agri-tech Career Conversion Programme (CCP), diploma or degree holders will undergo six months of training to become agri-tech specialists in roles such as farm managers, engineers, crop scientists, agronomists, or soil management experts, while non-PMETs can qualify for positions like supervisors, farm technicians, and operations executives after three months of training.[197]

Still, agricultural automation presents a double-edged sword.[198] While it increases productivity and efficiency, it also raises concerns about job displacement and widening inequalities if not implemented inclusively.[199] Policymakers must develop supportive legal frameworks, build infrastructure, and ensure equitable access to technology, especially for small-scale farmers.[200] These investments should be tailored to local contexts by addressing existing gaps in connectivity, education, and environmental priorities.[201]

One example of this transformation is the development of automated poultry farms, which integrate advanced technologies to optimize processes such as feeding, egg collection, and environmental control,

---

[196] The Business Times © SPH Media Limited. (2022b, January 22). Singapore's agri-tech sector will add 4,700 jobs by 2030; new scheme launched to retrain mid-career workers for industry. EDB Singapore. Retrieved April 10, 2025, from https://www.edb.gov.sg/en/business-insights/insights/singapore-s-agri-tech-sector-will-add-4700-jobs-by-2030-new-scheme-launched-to-retrain-mid-career-workers-for-industry.html.

[197] *Ibid.*

[198] Food and Agriculture Organization (FAO). (2022). The State of Food and Agriculture 2022 – Leveraging automation in agriculture for transforming agrifood systems. https://www.fao.org/3/cb9479en/cb9479en.pdf.

[199] *Ibid.*

[200] World Bank. (2019). Harvesting Prosperity: Technology and Productivity Growth in Agriculture. https://openknowledge.worldbank.org/entities/publication/acb134b2-3b82-5ccf-87a0-d6882a53d92e.

[201] McKinsey & Company. (2020). Agriculture's connected future: How technology can yield new growth. https://www.mckinsey.com/industries/agriculture/our-insights/agricultures-connected-future-how-technology-can-yield-new-growth.

resulting in greater efficiency and productivity.[202] By reducing reliance on manual labor, automated systems enable farmers to scale operations while maintaining consistent product quality.[203] These innovations extend beyond individual farms, as demonstrated by Singapore's Seng Choon egg farm, which significantly improved productivity through process automation supported by the Agri-Food and Veterinary Authority (AVA).[204] Seng Choon produces 625,000 eggs daily with just 100 workers by leveraging automation in key farm operations like feeding, temperature regulation, and waste management, significantly boosting efficiency and output.[205] Auto-egg inspection and crack detection involves a multi-stage process where eggs are first analyzed with spectrographic technology for dirt and internal abnormalities, then tested for cracks using advanced sound technology, and finally sanitized with ultraviolet light to minimize bacterial contamination.[206]

Artificial intelligence (AI) is revolutionizing sustainable farming by integrating technologies such as machine learning, robotics, and data analytics to enhance agricultural efficiency and productivity.[207] For instance, AI-driven platforms like Zordi, an agtech startup backed by Khosla Ventures, combine robotics and AI to automate greenhouse farming, optimizing conditions for crops like strawberries while reducing labor requirements.[208] These advancements enable precise monitoring and control of environmental factors, leading to increased yields and more sustainable farming practices.[209]

Integrating artificial intelligence (AI) into agricultural energy management offers both economic and environmental advantages, with

---

[202] Building Automated Poultry Farm Transforming from Manual to Automated with Automation Equipment. (n.d.). https://www.agico.com.cn/news/building-automated-poultry-farm.html.

[203] *Ibid.*

[204] Farm. (2024c, July 4). Default. https://www.sfa.gov.sg/fromSGtoSG/farms/farm/Detail/seng-choon.

[205] *Ibid.*

[206] Seng Choon Farm. (n.d.). https://www.sengchoonfarm.com/sub-page/technology.html.

[207] Bassett, A. (2023, November 14). Is AI the answer to sustainable farming? The Verge. https://www.theverge.com/2023/11/14/23950666/ai-sustainable-farming-machine-learning-agriculture.

[208] *Ibid.*

[209] *Ibid.*

AI reducing energy consumption in agriculture by up to 15%.[210] This reduction leads to significant operational cost savings and enhances the economic viability of farming operations.[211] AI enhances sustainability in greenhouses and vertical farms by optimizing energy use.[212]

In San Francisco, Plenty has built a fully automated indoor vertical farm using robotic arms and AI to manage stacked towers of crops with optimized lighting and climate control.[213] Plenty's automated farming system employs a coordinated array of robotic arms equipped with delicate gripping mechanisms that carefully remove seedlings from their trays and accurately position them along a linear metal planting track.[214] According to TechWire Asia, Plenty employs robotic systems to move crops within its vertical farms and uses artificial intelligence to regulate water, temperature, and lighting, enabling significantly higher yields within compact indoor spaces and addressing the global scarcity of land suitable for fresh produce cultivation.[215]

In commentary on agri-tech innovation, Rien Kamman notes that farmers already possess strong technical knowledge and experience, and that artificial intelligence should be used to equip them with better tools to improve efficiency and long-term success.[216]

Semiconductor sensors offer farmers advanced, durable, and energy-efficient tools to monitor animal health, welfare, and reproductive cycles,

---

[210] Simos, N. (2024, February 29). How AI is Powering Transformation in Agricultural Energy Management — AGRITECTURE. Agritecture. Retrieved March 2, 2025, from https://www.agritecture.com/blog/ai-in-agriculture-energy-management.

[211] *Ibid.*

[212] *Ibid.*

[213] Dua, S. (2023, December 29). Plenty's high-tech robot farm is transforming traditional agriculture. Interesting Engineering. https://interestingengineering.com/innovation/plentys-high-tech-robot-farm.

[214] *Ibid.*

[215] Kaur, D. (2020, December 31). AI and robot-run farms are transforming agriculture. TechWire Asia. https://techwireasia.com/2020/12/ai-and-robot-run-farms-are-transforming-agriculture/.

[216] Vedantam, K. and Vedantam, K. (2023, June 22). Here Are 3 Big Areas Where AI Is Cropping Up In Agtech. Crunchbase News. https://news.crunchbase.com/ai-robotics/agtech-indoor-farming-openai-plenty/.

enabling early issue detection and cost reduction.[217] Similarly, advanced crop monitoring technologies provide agronomists with real-time insights to enhance decision-making, supporting precision agriculture, early detection of issues, efficient resource use, sustainable practices, and optimized yields.[218] Among these technologies, AI-powered drones offer small and medium-scale farms precise, location-specific data and continuous updates, enabling timely interventions and driving increased adoption due to their immediate data access and detailed analysis.[219] In terms of drone types, fixed-wing drones are ideal for rapid, large-scale mapping over hundreds of hectares, while multirotor drones provide agility and precision for detailed imaging and localized treatments on smaller plots.[220]

Other innovators include Blue River Technology, whose mission is to revolutionize agriculture by developing intelligent machinery that enhances farming efficiency and sustainability.[221] They achieve this by integrating computer vision, machine learning, and robotics to create solutions like their "See & Spray" system, which precisely targets weeds, reducing herbicide usage by up to 77% and minimizing environmental impact.[222] Founded in 2011 and acquired by John Deere in 2017, Blue River Technology continues to innovate in agricultural robotics, aiming to make a tangible and positive impact on the planet.[223]

---

[217] Singer, P. (2024, February 15). Semiconductor innovations in the agriculture market. Semiconductor Digest. https://www.semiconductor-digest.com/semiconductor-innovations-in-the-agriculture-market/.

[218] Marinho, C. (2024, October 19). Smart Crop Monitoring: Boosting Agricultural Efficiency – Alteia. Alteia. https://alteia.com/resources/blog/crop-monitoring/.

[219] Abramov, M. (2025a, January 12). AI Drones inAgriculture:Transforming Crop Monitoring|KeyMakr. Keymakr. https://keymakr.com/blog/ai-drones-in-agriculture-transforming-crop-monitoring-and-precision-farming/.

[220] *Ibid.*

[221] Blue River Technology. (n.d.). Our mission is to create intelligent machinery that solves monumental challenges for our customers. Retrieved March 2, 2025, from https://www.bluerivertechnology.com/our-mission/.

[222] *Ibid.*

[223] *Ibid.*

Meanwhile, CropX, originally founded in New Zealand and now headquartered in Israel, is advancing soil sensor technologies and data platforms for efficient irrigation.[224] The company has expanded through several acquisitions, including CropMetrics in 2019, Regen in 2020, Dacom Farm Intelligence in 2021, Tule Technologies, and Green Brain, strengthening its precision irrigation and digital agriculture capabilities.[225] These acquisitions enhance CropX's ability to innovate in irrigation, water usage monitoring, disease control, and market expansion.[226]

Concluding Remarks

The exploration of Agriculture 4.0 presents a compelling vision for the future of food production, one that is both technologically advanced and environmentally responsive. The ability to decouple agriculture from traditional constraints such as climate variability, land scarcity, and seasonal limitations marks a significant advancement in sustainable farming practices. CEA's capacity to produce food in vertically integrated systems, as seen in innovations like Sky Greens and other urban farms in Singapore, illustrates the transformative potential of modern Agri-tech to enhance food security in densely populated and resource-constrained environments. These systems not only improve land-use efficiency but also drastically reduce water consumption and eliminate the need for chemical inputs, aligning well with global sustainability goals.

However, while these developments are promising, they are not without limitations. High capital investment requirements, intensive energy consumption, and the complexity of advanced systems may limit accessibility, particularly for smallholder farmers and communities in developing regions. Furthermore, the digital divide, lack of technical training, and inadequate infrastructure in many parts of the world pose

---

[224] CropX. (2025, February 17). About us – CROPX Digital Agronomy Platform and farm Management System. CropX Digital Agronomy Platform and Farm Management System. https://cropx.com/about-us/.

[225] *Ibid.*

[226] *Ibid.*

significant barriers to the equitable adoption of these innovations. This affirms the need for inclusive policies, targeted investment, and educational support to democratize access to these technologies. In addition, emphasis must be placed on ensuring that these systems are powered by renewable energy sources and embedded within circular economic frameworks to mitigate environmental impacts.

As the global population continues to rise and climate risks intensify, the role of advanced agricultural systems becomes increasingly vital. Strategic public–private partnerships, global knowledge exchange, and local capacity-building efforts are essential to scale these solutions sustainably. Moreover, fostering innovation hubs, supporting research and development, and encouraging youth participation in agri-tech fields can accelerate the transition to smarter food systems. Governments and international organizations must also provide regulatory support and financial incentives to reduce barriers to entry and de-risk adoption for farmers and investors alike. Initiatives such as training programs, micro-financing, and cooperative business models could play a key role in creating more inclusive and resilient agri-tech ecosystems.

Ultimately, while Agriculture 4.0 offers a powerful framework for addressing global food challenges, its success will depend on thoughtful integration, equitable access, and sustained cross-sector collaboration. These innovations must not only enhance productivity and resilience but also contribute to a more just, inclusive, and sustainable global food system, one that places technological innovation in the service of people, communities, and the planet. As we confront the intersecting crises of climate change, resource depletion, and global hunger, embracing such forward-looking agricultural strategies is not merely an option; it is an urgent necessity.

Looking ahead, the journey toward a technologically driven and ecologically sustainable food system requires continued commitment from all sectors of society. Future agricultural landscapes must be built on a foundation of accessibility, education, and ethical governance to ensure that innovation serves both people and the planet. Embracing the full potential of Agriculture 4.0 involves not only deploying cutting-edge technologies but also fostering inclusive ecosystems where smallholder farmers, urban communities, and young innovators can thrive. As new breakthroughs emerge, from AI-powered analytics to bioengineered crops and energy-efficient infrastructures, stakeholders must collaborate to create a resilient and regenerative agri-food system. Only through such collective effort can we achieve a future where food systems are not only productive and efficient but also equitable, adaptive, and aligned with planetary boundaries.

Global Mushroom Trends and Benefits

In countries with limited land, mushrooms offer a sustainable solution for agricultural production due to their ability to grow in confined spaces, such as vertical farms, indoor environments, or on agricultural by-products like straw or sawdust. They require minimal land area, making them ideal for urban farming or integration into existing food systems. Additionally, mushroom cultivation has a low environmental impact, using fewer resources like water and space compared to traditional crops, while providing a rich source of nutrition and potential income for small-scale farmers.

Mushrooms, the fruiting bodies of macroscopic fungi, have been esteemed in human diets for their culinary versatility, nutritional richness, and medicinal properties for centuries.[227] Ancient cultures, including the Egyptians, Greeks, Maya, and Aztecs, revered mushrooms for their mystical, hallucinogenic, and spiritual properties, often associating them with divinity, visions, and rituals.[228]

The Greek physician Hippocrates, around 450 BCE, recognized the amadou mushroom (*Fomes fomentarius*) for its anti-inflammatory properties and its use in cauterizing wounds.[229] Similarly, in the 5th century, the Chinese alchemist Tao Hongjing documented several medicinal mushrooms, including *lingzhi* (*Ganoderma lucidum*) and *zhu ling* (*dendropolyporus umbellatus*), some reportedly used by Shennong centuries earlier.[230]

---

[227] Valverde, M. E., Hernández-Pérez, T., and Paredes-López, O. (2015). Edible mushrooms: improving human health and promoting quality life. *International Journal of Microbiology*, 2015, 376387. https://doi.org/10.1155/2015/376387.

[228] Welle, D. (2022c, October 17). A brief cultural history of the mushroom. dw.com. https://www.dw.com/en/a-brief-cultural-history-of-the-mushroom/a-63461380.

[229] Stamets, P. and Zwickey, H. (2014). Medicinal mushrooms: Ancient remedies meet modern science. *Integrative Medicine (Encinitas, Calif.)*, 13(1), 46–47.

[230] *Ibid.*

Remarkably, Ötzi the Iceman, who lived nearly 5300 years ago, carried amadou and birch polypore mushrooms in a pouch, likely for medicinal purposes.[231] "The ancient society of the Iceman most likely already had a considerable knowledge about medical treatment," noted Albert Zink, head of the Eurac Research Institute for Mummy Studies.[232] This deep historical connection points to humanity's long-standing relationship with fungi, shaping both cultural history and dietary practices.[233]

Medicinal and Nutritional Significance

Ganoderma lucidum, known in Chinese as Lingzhi and often interpreted as "miraculous," "divine," or the "mushroom of immortality," is the most renowned and highly esteemed medicinal mushroom due to its long-standing use in traditional East Asian medicine for enhancing health and promoting longevity.[234] Anti-hypertensive medications have effectively reduced cardiovascular disease (CVD) risk from 55% to as low as 18%, and similar anti-hypertensive bioactive components such as polysaccharides, triterpenoids, ganoderic acids, ganoderic aldehydes, and ganoderic alcohols have been identified in both edible and therapeutic mushrooms, particularly in Ganoderma lucidum, whose angiotensin-converting enzyme (ACE) inhibitory peptides have shown promising results in lowering blood pressure.[235]

Mushrooms have been traditionally used across various cultures, and nutraceutical varieties are now recognized for their role in promoting holistic health and preventing disease, underscoring their potential value

---

[231] *Ibid.*

[232] Ghose, T. (2018, September 26). Ötzi the Iceman's Tattoos May Have Been a Primitive Form of Acupuncture. livescience.com. https://www.livescience.com/63682-otzi-ice-man-took-medical-treatment.html.

[233] *Ibid.*

[234] Diptayan Paul, Shreya Kundu, and Diptayan Paul. (2024, June). Exploring the therapeutic properties of chinese mushrooms with a focus on their anti-cancer effects: A systemic review. *Science Direct*. Retrieved April 10, 2025, from https://www.sciencedirect.com/science/article/pii/S2667142524000769.

[235] Teja K. S., Suruchi, R. U., Kumar M., Mohanty O., Roy J., Meshram S. Mushrooms: A potential option in the management of deficiency and diseases in humans. *Journal of Pure and Applied Microbiology*, 2023;17(2):749–760. DOI: 10.22207/JPAM.17.2.55.

in contemporary diets.[236] "We were surprised not to find any remains of meat in the Asturias Neanderthals, given that they were thought to be predominantly meat eaters. However, we have found evidence they enjoyed a varied diet including a wide range of plants. What's more, some of these plants may well have been cooked before being eaten," points out CSIC investigator, Antonio Rosas, who works in Spain's National Natural Science Museum.[237] The genomic analyses of dental plaque showed these Asturian Neanderthals ate wild mushrooms (*Schizophyllum commune*), pine nuts (*Pinus koraiensis*), moss (*Physcomitrella patens*), and western balsam-poplar (*Populus trichocarpa*).[238]

Mushrooms Culinary Diversity

Gourmet mushrooms celebrated worldwide for their unique flavor and status as a culinary wonder include over 2,000 species in nature, though only about 25 are widely consumed as food, with a few commercially cultivated.[239] Button mushrooms (*Agaricus bisporus*), the mildest-tasting variety, are the most common type found in grocery stores, making up about 90% of the mushrooms consumed in the US.[240]

Portobello mushrooms, with their large, meaty caps and robust texture, make a perfect beef substitute in vegan dishes, offering rich, earthy flavors and a satisfying chewiness ideal for grilling, roasting, or stuffing.[241] Tan-capped "baby bellas," a variety of Agaricus bisporus, are

---

[236] Hom-Singli Mayirnao, Karuna Sharma, Pooja Jangir, Surinder Kaur, Rupam Kapoor. (2024, June). Mushroom-derived nutraceuticals in the 21st century: An appraisal and future perspectives. *Science Direct*. Retrieved April 10, 2025, from https://www.science direct.com/science/article/pii/S2772566924000478.

[237] Milligan, M. (2017, June 9). Neanderthals at El Sidrón ate a diet of wild mushrooms, pine nuts and moss. HeritageDaily – Archaeology News. https://www.heritagedaily. com/2017/03/neanderthals-at-el-sidron-ate-a-diet-of-wild-mushrooms-pine-nuts-and-moss.

[238] *Ibid.*

[239] Valverde, M. E., Hernández-Pérez, T., and Paredes-López, O. (2015). *Op.cit.*

[240] Denenberg, Z. (2023b, March 21). 15 types of mushrooms and how to cook with them. Epicurious. https://www.epicurious.com/ingredients/types-of-mushrooms.

[241] Wysocarski, E. (2024c, September 23). Can mushrooms replace meat? 28 recipes that say yes. Olives for Dinner. https://olivesfordinner.com/mushroom-meat.

harvested later than button mushrooms, giving them a larger size, richer flavor, and heartier texture.[242]

Shiitake Mushroom (*Lentinula edodes*) is a highly valued and versatile mushroom, celebrated worldwide for its unique umami flavor and nutritional benefits in both culinary and medicinal traditions.[243] Oyster mushrooms (*Pleurotus ostreatus*), named for their oyster shell-like appearance, come in white, gray, and pink colors, with a fan-shaped structure and a mild, nutty flavor, making them ideal for sautéing, grilling, and stir-frying.[244]

Porcini mushrooms (*Boletus edulis*) have a rich, nutty flavor and are often featured in Italian dishes like risottos and pasta sauces.[245] White Enoki mushrooms (*Flammulina velutipes*), prized for their slender stems, delicate caps, crisp texture, and mild, slightly fruity flavor, are a nutritious staple in Asian cuisine, enhancing soups, stir-fries, and salads.[246]

The maitake mushroom, also called "hen-of-the-woods," "sheep head," and "dancing mushroom," is a vitamin D-rich fungus with an earthy aroma, a wholesome taste, and potential benefits for cancer support and bone health.[247] Lion's mane mushrooms (*Hericium erinaceus*), also called *hou tou gu* or *yamabushitake*, are large, white, shaggy fungi with culinary and medicinal uses in Asian countries like China, India, Japan, and Korea, renowned for their bioactive compounds that support brain, heart, and gut health.[248]

---

[242] Chappell, M. M. (2024, April 12). 12 types of mushrooms and how to use them. Forks Over Knives. https://www.forksoverknives.com/how-tos/mushrooms-types-how-to-use-them/#baby.

[243] Rootlab. (n.d.). Shiitake Mushrooms (Lentinula edodes) for sale | Rootlab. https://www.rootlab.com.au/mushroom-supplies/shiitake-mushrooms.

[244] WebstaurantStore. (2025b, January 22). Different types of mushrooms. https://www.webstaurantstore.com/article/928/types-of-mushrooms.html.

[245] Roach, C. (2025, February 11). Most common types of mushrooms. Variety Mode. https://www.varietymode.com/blogs/news/most-common-types-of-mushrooms.

[246] Out-Grow. (n.d.). Grow white enoki mushrooms – Liquid Culture syringe. Out Grow. https://www.out-grow.com/products/white-enoki-flammulina-velutipes.

[247] WebMD Editorial Contributor. (2024b, December 22). Maitake mushroom: Health benefits, nutrition, and uses. WebMD. https://www.webmd.com/diet/maitake-mushroom-health-benefits.

[248] Clt, E. J. M. R. (2024b, January 12). 9 Health benefits of Lion's Mane mushroom (Plus Side Effects). Healthline. https://www.healthline.com/nutrition/lions-mane-mushroom.

Sustainability and Market Trends

Mushrooms are an efficient and sustainable food source, requiring less land, energy, and water compared to traditional livestock.[249] Meat production has grown significantly in recent decades and is expected to continue expanding, but its high environmental impact due to land, water, and feed demands, as well as substantial greenhouse gas emissions that drive climate change, biodiversity loss, and nitrogen cycle disruptions, raises the question of whether mushrooms could effectively replace red meat to reduce these effects.[250] The fungi used to produce the product are cultivated on nutrient-rich food waste, enriching them with protein, iron, and amino acids, making them more nutritious than typical plant-based meat ingredients like peas, chickpeas, wheat gluten, and soy. [251]

The global mushroom market, driven by rising health consciousness and increasing demand for nutritional foods, is projected to grow at a CAGR of 6.74%, reaching 24.05 million tons by 2028 from 14.35 million tons in 2020 and 15.25 million tons in 2021, according to Fortune Business Insights.[252] Functional mushrooms like lion's mane, *reishi*, turkey tail, cordyceps, and *chaga* are a multi-billion-dollar wellness industry trend, touted for potential benefits such as disease prevention and cognitive enhancement, though further human studies are required.[253]

Innovations in Mushroom Farming

Innovative mushroom farming techniques, including biotechnology, vertical farming, and automation, are enhancing productivity, sustainability, and efficiency while addressing food security and environmental

---

[249] Pashaei, K. H. A., Irankhah, K., Namkhah, Z., and Sobhani, S. R. (2024). Edible mushrooms as an alternative to animal proteins for having a more sustainable diet: A review. *Journal of Health Population and Nutrition*, 43(1). https://doi.org/10.1186/s41043-024-00701-5.

[250] *Ibid.*

[251] Fungi-based protein healthier and greener alternative to plant-based meat. (2022, August 29). Corporate NTU. https://www.ntu.edu.sg/news/detail/growing-mushrooms-to-replace-meat.

[252] Fortune Business Insights. (2022, April 29). Mushroom market to reach 24.05 million tonnes by 2028. Retrieved April 12, 2025, from https://www.fortunebusinessinsights.com/press-release/mushroom-market-9301.

[253] Akhtar, A. (2022, April 7). 5 "functional" mushrooms the wellness industry is obsessed with, from lion's mane to turkey tail. Business Insider. https://www.businessinsider.com/what-functional-mushrooms-do-reishi-chaga-lions-mane-turkey-tail-2022-4.

challenges.[254] These advancements, such as controlled environments and precision farming, optimize resource use and enable year-round cultivation, paving the way for a more resilient and profitable industry.[255] Emerging trends and research in mushroom consumption highlight their expanding roles in health, industry, and environmental sustainability, leveraging biotechnological innovations and exploring non-food applications like biodegradable materials and bioremediation.[256]

Sustainable Mushroom Farming in Singapore

Due to Singapore's limited land area and tropical climate, the growth of indoor, controlled-environment mushroom cultivation has become increasingly vital.[257] This method allows for year-round production and vertical farming, optimizing the use of space while minimizing waste.[258] While mushroom farming has traditionally not been a major agricultural sector, it now offers a viable, space-efficient solution to food security and sustainability.[259]

---

[254] The Farming Insider. (2024a, November 28). Innovative techniques in modern mushroom farming. Retrieved April 12, 2025, from https://thefarminginsider.com/innovative-mushroom-farming-techniques/.

[255] *Ibid.*

[256] Ogwu, M. C., Kosoe, E. A., and Izah, S. C. (2025). Future Trends and research Directions: Emerging Uses and Trends in mushroom Consumption Utilization. In Reference series in phytochemistry (pp. 1–16). https://doi.org/10.1007/978-3-031-52642-8_45-1.

[257] Food for thought | giving shrimp, mushroom, and other farms an added boost. (2022, June 9). Default. https://www.sfa.gov.sg/food-for-thought/article/detail/giving-shrimp-mushroom-and-other-farms-an-added-boost?trk=organization_guest_main-feed-card_feed-article-content.

[258] Food for Thought | Farming in unusual spaces. (2018, January 1). Default. https://www.sfa.gov.sg/food-for-thought/article/detail/farming-in-unusual-spaces.

[259] Somasundaram Jayaraman, Brijesh Yadav, Ram C. Dalal, Anandkumar Naorem, Nishant K. Sinha, Ch Srinivasa Rao, Y.P. Dang, A.K. Patra, S.P. Datta, A. Subba Rao, Mushroom farming: A review. Focusing on soil health, nutritional security and environmental sustainability. *Farming System*, Volume 2, Issue 3, 2024, 100098, ISSN 2949-9119, https://doi.org/10.1016/j.farsys.2024.100098.

A range of mushroom species, including oyster and shiitake, are cultivated to provide fresh, locally grown produce.[260] Technological innovation plays a critical role in Singapore's mushroom cultivation efforts.[261] Vertical farming systems, stacked mushroom bags, and advanced climate controls create ideal growing conditions.[262] These systems, which often include automated misting, effectively maintain the optimal humidity level for mushroom cultivation, reducing manual labor and optimizing resource utilization to enhance productivity.[263]

Ongoing research into substrate development and nutrient mixes aims to enhance mushroom growth and yield.[264] By leveraging technology, Singapore's farmers can cultivate high-quality mushrooms while minimizing their environmental impact.[265] The use of recycled materials, such as coffee grounds and agricultural waste, for substrate preparation supports a circular economy.[266] Oyster mushrooms are particularly

---

[260] Min, C. H. (2025, February 12). We tried eating mostly Singapore-grown food for a week – here's what happened. CNA. https://www.channelnewsasia.com/singapore/30-30-local-homegrown-food-farm-table-produce-4917766.

[261] *Fogo-ing leisure to grow mushrooms in bedroom.* (2023, December 20). *Singapore Management University News.* https://news.smu.edu.sg/news/2023/12/20/fogo-ing-leisure-grow-mushrooms-bedroom.

[262] Research, U. M. (2025, January 20). Transforming agriculture: The rise of urban mushroom farming. https://www.linkedin.com/pulse/transforming-agriculture-rise-urban-mushroom-farming-ldsjf/.

[263] Habibuddin, J., Muchtar, T., Azis, M. F., and Sasmita, M. R. (2023). Automatic mist sprayer for oyster mushroom cultivation using ThingsPeak IoT platform and R Analytical tool. *MOTIVECTION Journal of Mechanical Electrical and Industrial Engineering*, 5(2), 365–374. https://doi.org/10.46574/motivection.v5i2.241.

[264] Muswati, C., Simango, K., Tapfumaneyi, L., Mutetwa, M., and Ngezimana, W. (2021). The Effects of Different Substrate Combinations on Growth and Yield of Oyster Mushroom (Pleurotus ostreatus). *International Journal of Agronomy*, 2021, 1–10. https://doi.org/10.1155/2021/9962285.

[265] Metteo, R. (2024, June 28). Looking for cheaper mushrooms? You may soon get them from Singapore's newest farm that taps AI. CNA. https://www.channelnewsasia.com/singapore/mushroom-farms-artificial-intelligence-automation-cheaper-better-quality-yield-4442591.

[266] Fabio De Felice, Mizna Rehman, Antonella Petrillo, Miguel Angel Ortiz Barrios, Ilaria Baffo, Integrating IoT and circular economy in textile supply chains: A closed-loop model for sustainable production using recycled PET and spent coffee grounds. *Journal of Cleaner Production*, 2025, 145226, ISSN 0959-6526, https://doi.org/10.1016/j.jclepro.2025.145226.

well-suited to Singapore's climate due to their fast growth cycle and ease of cultivation, making them a popular choice.[267] Both common gray and visually striking pink oyster mushrooms are grown, adding diversity in flavor and appearance.[268]

Vertical farming systems are key to overcoming Singapore's land scarcity and tropical conditions.[269] Singapore's newest mushroom farm, operated by Finc Bio-Tech Mushroom, is set to revolutionize local agriculture by employing artificial intelligence (AI) and automation to produce white button mushrooms more efficiently and affordably.[270] Located at Sungei Tengah Close on a plot the size of three football fields, the vertical farm aims to supply over a third of Singapore's white button mushroom demand at approximately S$7.50 per kg, about half the current market price.[271]

By leveraging AI-driven systems to monitor and predict mushroom growth daily, the farm anticipates achieving a 60-day cultivation cycle with yields up to 2,500 tons annually, more than 50% higher than traditional methods.[272] These technologies enable precise monitoring and daily predictions of mushroom growth metrics like size and weight, optimizing growing conditions to enhance yield and quality.[273] Singapore's mushroom farms are utilizing real-time data analytics to monitor critical environmental parameters, enabling precise adjustments that optimize both yield and quality.[274]

---

[267] David, S. (2016, April 27). Local produce you didn't know could be grown in Singapore. MICHELIN Guide. https://guide.michelin.com/sg/en/article/dining-out/local-produce-you-didn-t-know-could-be-grown-in-singapore.

[268] GroCycle. (2019). *How to grow oyster mushrooms: The ultimate step-by-step guide.* https://grocycle.com/wp-content/uploads/2019/06/How-To-Grow-Oyster-Mushrooms-The-Ultimate-Step-By-Step-Guide-Ebook.pdf.

[269] Metteo, R. (2024c, June 28). Looking for cheaper mushrooms? You may soon get them from Singapore's newest farm that taps AI. CNA. https://www.channelnewsasia.com/singapore/mushroom-farms-artificial-intelligence-automation-cheaper-better-quality-yield-4442591.

[270] *Ibid.*

[271] *Ibid.*

[272] *Ibid.*

[273] New mushroom farm to be realized in Singapore. (2024, July 1). https://www.vertical farmdaily.com/article/9640326/new-mushroom-farm-to-be-realized-in-singapore/.

[274] CNA. (2024, June 27). Singapore's newest mushroom farm to use AI, automation to boost crop yield [Video]. YouTube. https://www.youtube.com/watch?v=pzjf4mnlNL0.

The advancement in mushroom cultivation systems reflects the global move toward precision agriculture, leveraging data-driven decisions to optimize resource use efficiently.[275] Concurrently, research is being directed toward optimizing substrate composition through the use of sustainable materials such as coffee grounds and agricultural waste.[276] These substrates undergo advanced mixing and pasteurization processes to eliminate contaminants and promote optimal fungal growth.[277] Collectively, these innovations position Singapore's mushroom farms at the forefront of sustainable and technologically advanced agriculture, setting a benchmark for efficiency, resource optimization, and environmental responsibility in urban farming.[278]

The market for mushrooms in Singapore is growing steadily, fueled by an increasing awareness of their health benefits and a preference for locally sourced produce.[279] Mushrooms are rich in protein, fiber, vitamins, and minerals, making them a popular choice for health-conscious consumers.[280] This demand has spurred an expansion in the market, with restaurants, cafes, and food stalls increasingly incorporating mushrooms into their menus.[281] Mushrooms, like golden, oyster, shiitake, and white button, have become staples in local dishes, contributing to the trend of

---

[275] Rukhiran, M., Sutanthavibul, C., Boonsong, S., and Netinant, P. (2023). IoT-based mushroom cultivation system with solar renewable energy integration: Assessing the sustainable impact of the yield and quality. *Sustainability*, 15(18), 13968. https://doi. org/10.3390/su151813968.

[276] Barta, D. G., Simion, I., Tiuc, A. E., and Vasile, O. (2024). Mycelium-based composites as a sustainable solution for waste management and circular economy. *Materials (Basel, Switzerland)*, 17(2), 404. https://doi.org/10.3390/ma17020404.

[277] Pasteurization and sterilization. (n.d.). La Mycosphère. Retrieved April 12, 2025, from https://lamycosphere.com/en/pages/pasteurization-and-sterilization.

[278] Mc-Author and Mc-Author. (2024, August 8). Explore sustainable urban agriculture in Singapore. MC Solutions – System Development to Facilitate GX (Green Business) – Web Application Development – Web System Development – Ordering System Development. https://mcsolutions.vn/sustainable-urban-agriculture-in-singapore-latest-update-2024/.

[279] Technavio. (2020, September 20). TechNavio Newsroom. Technavio. https://newsroom. technavio.org/singapore-fresh-mushroom-market-industry-analysis.

[280] Valverde, M. E., Hernández-Pérez, T., and Paredes-López, O. (2015). *Op.cit.*

[281] The new "Fun Guys": 'Shrooms, hot honey & Southeast Asian flavors to spice up restaurant menus in 20. (n.d.). NRA. https://restaurant.org/research-and-media/media/ press-releases/mushrooms-hot-honey-southeast-asian-flavors-spice-up-restaurant-menus-in-2025/.

healthier, more sustainable eating.[282] King oyster mushrooms are often favored for their versatility and affordability, while shiitake mushrooms, with their umami flavor, are popular in soups and stir-fries.[283] Button mushrooms, though less popular, are commonly used in salads and pizzas.[284] Specialty mushrooms, including shiitake, oyster, and lion's mane, are gaining popularity, with US sales reaching \$87.3 million in the 2021 to 2022 season, representing a 32% increase from its previous year.[285]

The increasing variety of mushrooms shows the evolving culinary preferences of Asian consumers, alongside a growing demand for healthier and more sustainable food options.[286] Distribution channels are efficient, with local farms supplying directly to supermarkets, restaurants, and food service providers.[287] Some farms also operate retail outlets, allowing consumers to buy mushrooms directly from the source.[288] The "SG Fresh Produce" logo promotes local produce, encouraging consumers to support homegrown goods and even try their hand at home gardening.[289] According to Professor William Chen, director of Nanyang Technological University's Food Science and Technology Programme, while consumers tend to be conservative and hesitant to step outside their comfort zones, locally grown produce often surpasses imported

---

[282] Lamontagne, J. (2024, March 17). Exploring Consumer Preferences: The Top-Selling Mushroom Varieties. Le Petit Champi. https://petitchampi.com/blogs/news/exploring-consumer-preferences-the-top-selling-mushroom-varieties.

[283] Wysocarski, E. (2024, September 23). Can mushrooms replace meat? 28 recipes that say yes. Olives for Dinner. https://olivesfordinner.com/mushroom-meat/.

[284] Denenberg, Z. (2023, March 21). 15 types of mushrooms and how to cook with them. Epicurious. https://www.epicurious.com/ingredients/types-of-mushrooms.

[285] Boosting the quality and safety of specialty mushrooms [Fact Sheet] | 1 extension. (2023, July 14). Extension. https://extension.unh.edu/resource/boosting-quality-safety-specialty-mushrooms-fact-sheet.

[286] Mushrooms, mushrooms! (n.d.-c). https://www.healthhub.sg/live-healthy/mushrooms-mushrooms.

[287] Metteo, R. (2024b, June 28). Looking for cheaper mushrooms? You may soon get them from Singapore's newest farm that taps AI. CNA. https://www.channelnewsasia.com/singapore/mushroom-farms-artificial-intelligence-automation-cheaper-better-quality-yield-4442591.

[288] Our SG food story. (2025, March 13). Default. https://www.sfa.gov.sg/fromSGtoSG/our-sg-food-story.

[289] *Ibid.*

alternatives in both freshness and nutritional value due to shorter harvesting periods, reduced cold storage, and minimal transit time.[290]

However, several farms in Singapore are also facing several challenges.[291] The initial investment for setting up indoor farms with the necessary technology, such as climate control, LED lighting, and automated systems, can be prohibitively high, especially for smaller-scale operations.[292] This financial burden is a significant barrier for new entrants into the market.[293] Furthermore, the technical expertise required to manage complex mushroom farms, such as knowledge of fungal biology, substrate preparation, and climate control, can limit the number of qualified farmers.[294] More training and educational programs are needed to bridge this skills gap and improve the efficiency of mushroom farming.[295]

Disease and pest management remain ongoing challenges in mushroom farming, particularly when organic substrates are used.[296] Even in controlled environments, fungal infections and pest infestations can spread rapidly if not properly managed, requiring strict hygiene protocols and targeted pest control measures.[297] Implementing effective strategies, such as Integrated Pest Management (IPM), involves combining approaches such as biological control using beneficial insects, along with regular monitoring to detect issues early.[298] Continued research and

---

[290] Tham, D. (2024, July 3). IN FOCUS: Will Singapore meet its 30 by 30 food sustainability goal? CNA. https://www.channelnewsasia.com/singapore/vegetables-seafood-singapore-food-price-cost-farmers-security-4442626.

[291] Loh, R. (2024b, October 21). With several farms closing or struggling to break even, what is the future for agriculture in Singapore? CNA. https://www.channelnewsasia.com/today/big-read/high-tech-low-returns-farming-4684566.

[292] Irfan, S. (2023, November 23). Vertical farming startup: A guide to your funding strategies. Vertical Farms Ltd. https://vertical.mt/vertical-farming-startup.

[293] *Ibid.*

[294] Fabricio Rocha Vieira, John Andrew Pecchia, Fungal community assembly during a high-temperature composting under different pasteurization regimes used to elaborate the Agaricus bisporus substrate. *Fungal Biology*, Volume 125, Issue 10, 2021, Pages 826–833, ISSN 1878-6146, https://doi.org/10.1016/j.funbio.2021.05.004.

[295] Gabriel, S. (2025, March 31). Community Mushroom Educator Program – Cornell Small Farms. Cornell Small Farms. https://smallfarms.cornell.edu/projects/mushrooms/cme/.

[296] The Farming Insider. (2024, December 17). Pest and disease management in mushroom farming. https://thefarminginsider.com/pest-disease-management-mushroom-farming/.

[297] *Ibid.*

[298] *Ibid.*

development play a vital role in overcoming these challenges and enhancing the overall efficiency and sustainability of mushroom cultivation.[299]

In addressing these challenges, farms like Golden Cap Farm demonstrate how effective management and innovation can support sustainable mushroom cultivation in an urban city like Singapore.[300] Founded by Benedict Yeo, Golden Cap Farm is a successful urban mushroom farm that cultivates gourmet varieties like Lion's Mane and partners with local businesses such as Rustica Café promote a thriving farm-to-table ecosystem.[301] By utilizing indoor farming setups, the farm not only maximizes space efficiency and minimizes environmental impact but also maintains controlled conditions that help prevent disease outbreaks and ensure consistent, year-round production regardless of external weather conditions.[302]

The sustainability of mushroom farming is further strengthened by implementing resource conservation practices.[303] Major on-farm water uses include preparing and sterilizing growing media, misting during the cultivation cycle, cleaning floors, and meeting the daily water needs of farm workers.[304] Mushrooms require fewer chemicals than conventional crops, contributing to lower environmental impact.[305] While the overall sustainability picture is positive, there are still areas for improvement,

---

[299] Li, H. and Akenroye, T. (2025, February 26). Supply chain optimization could boost vertical farming. Here's how. World Economic Forum. Retrieved March 16, 2025, from https://www.weforum.org/stories/2025/02/supply-chain-optimization-could-boost-vertical-farming/.

[300] Tay, L. (2025, February 26). Innovation with Purpose – A Singapore Farm-to-Table Story. Ieatishootipost. https://ieatishootipost.sg/innovation-with-purpose-a-singapore-farm-to-table-story/.

[301] *Ibid.*

[302] The advantages and challenges of vertical farming. (2024, June 26). https://www.vertical farmdaily.com/article/9638755/the-advantages-and-challenges-of-vertical-farming/.

[303] Ahmed, H., Abolore, R. S., Jaiswal, S., and Jaiswal, A. K. (2024). Toward circular economy: Potentials of spent coffee grounds in bioproducts and chemical production. *Biomass*, 4(2), 286–312. https://doi.org/10.3390/biomass4020014.

[304] Liyanage, Wasantha, De Silva, Siriwardana, and Atapattu, Mahinda. (2023). The water footprint of Oyster mushroom (Pleurotus ostreatus) cultivation under small-scale polybag farming conditions in Sri Lanka. *Journal of Agricultural Sciences – Sri Lanka*. 18. 172–182. 10.4038/jas.v18i2.10251.

[305] Robinson, N. (2024, August 15). Why Mushrooms are the future of sustainable agriculture. R&R Cultivation. https://rrcultivation.com/blogs/mn/why-mushrooms-are-the-future-of-sustainable-agriculture.

particularly regarding the energy consumption of indoor climate control systems.[306] Farms are exploring alternative energy sources, such as solar power, to reduce their carbon footprint further.[307] Additionally, the environmental impact of transporting certain substrate materials, like sawdust, is a consideration that must be addressed.[308]

Ongoing research aims to improve efficiency across all stages of the mushroom cultivation process, from substrate preparation to final harvest, ensuring that Singapore's mushroom farms continue to grow sustainably and meet increasing market demand.[309]

Gourmet Mushroom Cultivation

Singapore has a tropical climate with abundant rainfall, consistently high temperatures, and elevated humidity throughout the year due to its proximity to the equator.[310] Temperature and relative humidity are two climate variables that remain relatively stable, unlike temperate regions, where fluctuations occur more frequently across seasons and even within a single day.[311]

The daily minimum temperature in Singapore rarely drops below 23–25°C at night, while the maximum temperature seldom exceeds 31–33°C during the day.[312] The average monthly temperature peaks in May at 28.6°C and is lowest in December and January at around 26.8°C.[313]

---

[306] *Ibid.*

[307] The benefits of solar power. (n.d.). https://www.tech.gov.sg/media/technews/benefits-of-solar-power.

[308] Kayode Adesina Adegoke, Oreoluwa Ololade Adesina, Omolabake Abiodun Okon-Akan, *et al.* (2022). *Sawdust-biomass based materials for sequestration of organic and inorganic pollutants and potential for engineering applications. Current Research in Green and Sustainable Chemistry,* 5, 100274. https://doi.org/10.1016/j.crgsc.2022.100274.

[309] BIG feature: Fogo Fungi – From mushroom enthusiasts to sustainable tech innovators | Institute of Innovation & Entrepreneurship. (n.d.-d). Institute of Innovation & Entrepreneurship. https://iie.smu.edu.sg/node/9721.

[310] World Bank Climate Change Knowledge Portal. (n.d.). https://climateknowledgeportal.worldbank.org/country/singapore/climate-data-historical.

[311] Climate of Singapore |. (n.d.). http://www.weather.gov.sg/climate-climate-of-singapore.

[312] *Ibid.*

[313] *Ibid.*

This tropical environment is ideal for cultivating a wide variety of mushrooms, including popular species such as abalone mushrooms and golden oyster mushrooms, as well as lesser-known types like black fungus and pink oyster mushrooms, many of which can thrive in semi-indoor setups that provide adequate humidity, ventilation, and shade.[314] These mushrooms grow in Singapore's warm, humid conditions when cultivated under proper management.[315] Kin Yan Agrotech, the country's largest producer of mushrooms and wheatgrass since its establishment in 1997, grows these species as part of an economically viable and environmentally friendly business model.[316] The farm produces about 150 tons of mushrooms annually using spawn bags filled with agricultural byproducts such as rice husks, sugarcane, corn, and wood dust.[317] Kin Yan's Senior Farm Manager, See Jen Chuan, explains that wheatgrass is typically harvested around the seventh day when its nutritional value is highest, and its rapid growth results in minimal pest issues, eliminating the need for pesticides.[318]

Singapore's pioneering farming techniques have established the country as a center for sustainable and productive mushroom growing, utilizing various farming methods and capitalizing on its tropical climate.[319] Most of the outdoor mushroom fruiting chambers are placed in a shady location with good drainage, built with sturdy frames using materials like PVC pipes, wood, or metal, and equipped with shelves for the substrate.[320] The frame is covered with plastic sheeting or greenhouse plastic to maintain humidity, and has a waterproof, angled roof made of corrugated plastic or metal sheets for water to run off.[321] Some mushroom growing setups include a misting system to maintain humidity when growing certain mushrooms, keeping levels between 80% and 95% using

---

[314] N. F. (2021, June 25). Made in Singapore: Singapore's Largest Mushroom Farm. YouTube. https://www.youtube.com/watch?v=QB0RQ1NR_hQ.

[315] *Ibid.*

[316] *Ibid.*

[317] Singapore Economic Development Board. (2021, November 14). *Tech farming bears fruit in Singapore.* https://www.edb.gov.sg/en/business-insights/insights/tech-farming-bears-fruit-in-singapore.html.

[318] *Ibid.*

[319] N. F. (2021, June 25). *Op.cit.*

[320] *Ibid.*

[321] *Ibid.*

a humidifier, and misting regularly to maintain moisture.[322] Singapore's naturally high relative humidity, averaging around 82%, combined with frequent rainfall, provides an excellent environment for fungi to flourish.[323]

Certain companies manage the entire mushroom production cycle indoors, starting from controlled laboratory environments.[324] This allows them to customize the mushrooms according to specific tastes and sizes. Internal mycological laboratories are used to develop mushroom strains with high precision.[325] For instance, Bewilder, a mycotechnology firm established in Singapore, has an innovative method for cultivating fungi that commences with a meticulous process that entails the precise dissection of a mushroom and the isolation of a section of its inner flesh.[326] The chosen section is carefully placed into a sterile petri dish that has been extensively coated with a nutrient-rich agar medium.[327] The agar medium is specifically formulated to provide the required nutrients for the mycelium to grow and multiply.[328]

Other businesses take a streamlined approach by purchasing pre-colonized substrate blocks from international suppliers.[329] Instead of producing their own substrate, they procure mushroom substrate blocks from international suppliers.[330] These blocks already contain mycelium and allow businesses to focus primarily on the fruiting phase.[331]

---

[322] Wiberg, M. (2025, March 30). How to grow maitake mushrooms. Out Grow. https://www.out-grow.com/blogs/growing-mushrooms/how-to-grow-maitake-mushrooms.

[323] Xuan, L. W. (2024, January 30). Mould cases on the rise during rainy season. The Straits Times. https://www.straitstimes.com/singapore/mould-cases-on-the-rise-during-rainy-season.

[324] Fogo Fungi. (2025, August 4). *From mushroom enthusiasts to sustainable tech innovators – SMU BIG feature.* https://fogofungi.com/blogs/news/fogo-fungi-from-mushroom-enthusiasts-to-sustainable-tech-innovators-smu-big-feature.

[325] Southwest Mushrooms. (2021, February 12). A Mushroom Farm Laboratory | Southwest Mushrooms [Video]. YouTube. https://www.youtube.com/watch?v=W86NQf2ir7A.

[326] *Ibid.*

[327] Begum, S. (2022, June 16). New facility to produce gourmet mushrooms that can be customised, starting from petri dish. The Straits Times. https://www.straitstimes.com/singapore/environment/new-facility-to-produce-gourmet-mushrooms-that-can-be-customised-starting-from-a-petri-dish.

[328] *Ibid.*

[329] *Ibid.*

[330] How Chinese substrate produces "Product of USA" mushrooms. (2021, May 27). Far West Fungi. https://farwestfungi.com/blogs/news/chinese-substrate.

[331] *Ibid.*

By focusing primarily on the critical fruiting phase, these businesses can allocate resources to creating the optimal conditions that are essential for the mushroom's growth and the development of its fruiting bodies.[332] While this approach facilitates accelerated production and allows these businesses to meet market demands more efficiently, it does come with a trade-off.[333] This method accelerates production and enhances efficiency, though it sacrifices some control over the mushroom strain and overall product quality.[334]

Gourmet mushrooms are gaining popularity due to growing interest in plant-based diets and a desire for diverse culinary experiences.[335] Their unique flavors, textures, and nutritional benefits make them attractive meat alternatives, particularly for vegans.[336] The trend aligns with the rise of sustainable and locally sourced food movements.[337] Gourmet mushrooms can be cultivated in controlled environments using organic materials such as hardwood pellets and soybean hulls, appealing to environmentally conscious consumers.[338] This trend is consistent with customer preferences for ecologically conscious and ethically sourced meals, indicating a desire to understand the sources and manufacturing processes of their food.[339]

Singapore is promoting healthier eating habits by reducing sugar and sodium levels in hawker center foods.[340] This initiative aims to tackle chronic diseases like diabetes and hypertension by encouraging food vendors to use healthier ingredients and cooking methods.[341] Mushrooms can play a role in this initiative due to their nutritional

---

[332] *Ibid.*

[333] *Ibid.*

[334] *Ibid.*

[335] *Ibid.*

[336] Michlewicz, C. (2023b, December 5). The Growing Popularity of Gourmet Mushrooms Explained. Hydroponic Container Farms and Mushroom Farms – Farmbox Foods. https://farmboxfoods.com/the-growing-popularity-of-gourmet-mushrooms-explained/.

[337] *Ibid.*

[338] *Ibid.*

[339] *Ibid.*

[340] *Ibid.*

[341] Goh, C. (2022, October 9). The Big Read: As health consciousness rises, can a nation of foodies live with less sugar, salt and all things nice? CNA. https://www.channelnewsasia.com/today/big-read/healthier-sg-less-sugar-salt-sodium-diet-hawker-centres-nutrition-2996756.

benefits and versatility.[342] They are low in calories and can be used to enhance the flavor and texture of dishes without relying heavily on salt or sugar.[343] For instance, mushrooms can be added to soups, stir-fries, and even meat dishes to provide umami flavor, which can reduce the need for added sodium.[344] Additionally, they are rich in vitamins, minerals, and dietary fiber, contributing to a balanced diet.[345] Emerging research also highlights the potential of mushrooms to support immune function and reduce inflammation, further reinforcing their value in health-focused diets.[346] Their ease of preparation and compatibility with local culinary practices make them an accessible ingredient for hawker stalls.[347] As demand for nutritious and flavorful food options grows, mushrooms offer a sustainable solution that aligns with Singapore's public health priorities.[348] By incorporating more mushrooms into Singapore hawker center meals, vendors can offer healthier options that align with Singapore's dietary guidelines, supporting public health goals in the long term.[349]

Looking back at the transformative era of the 1980s, Dr. Tan Kok Kheng, a visionary leader of the Everbloom group, significantly impacted the agricultural industry through his pioneering work in mushroom farming.[350] As founder and CEO of MycoBiotech Inc., he led the company from its humble beginnings to prominence, focusing on innovative ideas.[351] Dr. Tan's leadership saw the development of facilities and the transition to commercial-scale production with a strong emphasis on quality and innovation.[352] One of Dr. Tan's notable achievements was securing a U.S. patent (US4987698A) in 1991 for a novel mushroom cultivation method.[353]

---

[342] *Ibid.*

[343] *Ibid.*

[344] *Ibid.*

[345] *Ibid.*

[346] UCLA Health. (n.d.). 7 health benefits of mushrooms. https://www.uclahealth.org/news/article/7-health-benefits-of-mushrooms.

[347] Healthy Hawker food. (n.d.). https://www.healthhub.sg/live-healthy/healthyhawkerfood.

[348] *Ibid.*

[349] *Ibid.*

[350] Tan, K. K. and Ltd, E. M. P. (1986, June 3). US4987698A – Mushroom cultivation – Google Patents. https://patents.google.com/patent/US4987698A/en.

[351] *Ibid.*

[352] *Ibid.*

[353] *Ibid.*

The method involved combining fungal spawn with hydrated, sanitized comminuted wood and applying thermal shock to optimize fungal growth and fruiting.[354] This technique not only increased yield and shortened cultivation cycles but also offered a sustainable and cost-effective alternative to traditional methods by repurposing waste materials.[355]

SPORE Garden, which specializes in mushroom production, primarily relies on automation to streamline operations and reduce unnecessary expenses.[356] The company is known for cultivating Lion's Mane, Yanagi Matsutake, and Pink Oyster mushrooms; these specific varieties are hard to find in Singapore because of their fragile characteristics and the complexities associated with their transportation.[357]

Fogo Fungi, the start-up, officially incorporated in May 2023, quickly outgrew its initial space and plans to move to a 4,000 square foot facility in Tuas by the end of 2023, enabling it to produce up to two tons of mushrooms per month.[358] The company's innovative approach and rapid growth highlight the potential of urban farming and sustainable food production in Singapore.[359] Mushrooms are nature's decomposers, and what Fogo Fungi is doing is recreating nature's process on an indoor scale, using mushrooms to upcycle food waste like spent beer grains, coffee waste, and okara to cultivate gourmet mushrooms.[360] This practice helps Fogo Fungi to further close the loop of the food system and contribute to a circular economy.[361] Mushrooms are highly efficient waste-to-value transformers, with a biological efficiency of 100–200% depending on

---

[354] *Ibid.*

[355] *Ibid.*

[356] S. G. (2023, April 30). CNA Growing Wild Ep4: Spore Gardens 1/2. YouTube. https://www.youtube.com/watch?v=MiZMSRubLf8.

[357] *Ibid.*

[358] Goh, T. (2023, December 19). Starting a start-up: From hotel manager to growing mushrooms in a spare bedroom. The Straits Times. https://www.straitstimes.com/business/companies-markets/starting-a-start-up-from-hotel-manager-to-growing-mushrooms-in-a-spare-bedroom.

[359] *Ibid.*

[360] BIG FEATURE: FOGO FUNGI – FROM MUSHROOM ENTHUSIASTS TO SUSTAINABLE TECH INNOVATORS. (2024, February 29). Singapore Management University – Institute of Innovation & Entrepreneurship. Retrieved June 13, 2024, from https://iie.smu.edu.sg/node/9721.

[361] *Ibid.*

the species, enabling at least 1 kg of fresh mushrooms to be grown from every 1 kg of waste material.[362]

The integration of the Internet of Things (IoT) into mushroom production provides growers with insights and control over the growth environment.[363] IoT sensors are placed within the substrate bed to collect real-time data on crucial factors, enabling accurate monitoring and regulation of conditions.[364] This data is transmitted to a centralized system for decision-making, allowing automated control systems to dynamically adjust ventilation, heating, and cooling based on predefined parameters.[365] The ability to remotely monitor and manage the growing process further optimizes yield and efficiency while minimizing resource waste.[366]

The technological synergy is mostly based on the interaction of AI algorithms and sensors.[367] While sensors serve as the eyes and ears of the operation, gathering real-time information about the physical environment, AI acts as the brain, interpreting and analyzing this data to extract actionable insights and drive informed decision-making.[368] AI holds the importance of predictive analytics, enabling growers to anticipate future trends and make proactive adjustments to their cultivation strategies.[369] By using historical data and real-time observations, AI algorithms can forecast optimal conditions for mushroom growth, preemptively mitigate threats, and optimize resource allocation for enhanced efficiency and productivity.[370]

The use of AI in mushroom cultivation represents a fundamental change in agricultural practices, initiating a new era of precision farming

---

[362] *Ibid.*

[363] United States Department of Agriculture, National Institute of Food and Agriculture. (n.d.). *IoT-based smart automated mushroom production system (Project No. 1026673).* https://training-portal.nifa.usda.gov/web/crisprojectpages/1026673-iot-based-smart-automated-mushroom-production-system.html.

[364] *Ibid.*

[365] *Ibid.*

[366] *Ibid.*

[367] *Ibid.*

[368] Sheldon, R. (2022, August 16). Sensor. WhatIs. https://www.techtarget.com/whatis/definition/sensor.

[369] *Ibid.*

[370] Beasley, K. (2023, October 5). Unlocking The Power Of Predictive Analytics With AI. Forbes. https://www.forbes.com/councils/forbestechcouncil/2021/08/11/unlocking-the-power-of-predictive-analytics-with-ai/.

and data-driven decision-making.[371] As technology continues to advance, the possibilities for innovation and optimization in mushroom cultivation are virtually limitless, promising a brighter and more sustainable future for the agricultural industry.[372] Artificial intelligence (AI) and the Internet of Things (IoT) can be leveraged to monitor and optimize the growth stages of the rare Lion's Mane mushroom, enhancing both its productivity and quality.[373]

Artificial intelligence systems can analyze sensor data and historical records in order to predict the optimal parameter changes that promote fruit development.[374] The variables under consideration include things like the type of fungus, the state of the weather now, and the stage at which the fungi are developing.[375] Artificial intelligence has the means to improve fruit production efficiency and yields by continuously learning new things and improving recommendations.[376] Building an artificial intelligence system that uses cameras to monitor and assess the state of mushrooms is possible by utilizing image recognition technology.[377] The system can identify early indicators of contamination, which facilitates the prompt execution of corrective measures.[378]

Analysis of images captured by the cameras is conducted by the AI system, which detects preliminary signs of contamination in the mushrooms.[379] This enables timely intervention, which must avert significant crop losses.[380] Elevation in carbon dioxide concentrations is detected by

---

[371] *Ibid.*

[372] Mycionics Inc. (2022, December 11). Revolutionize your Mushroom Farm using Robotic Harvesting | Mycionics Inc. [Video]. YouTube. https://www.youtube.com/watch?v=npGkqo-wewQ.

[373] *Ibid.*

[374] *Ibid.*

[375] *Ibid.*

[376] *Ibid.*

[377] *Ibid.*

[378] Leichtmann, B., Hinterreiter, A., Humer, C., Streit, M., and Mara, M. (2023). Explainable artificial intelligence improves human decision-making: Results from a mushroom picking experiment at a public art festival. *International Journal of Human-computer Interaction*, 1–18. https://doi.org/10.1080/10447318.2023.2221605.

[379] *Ibid.*

[380] GigE Camera Supports AgTech Study of Mushroom Cultivation in Greenhouses. (2023, January 4). Vision Systems Design. https://www.vision-systems.com/cameras-accessories/article/14287799/gige-camera-supports-agtech-study-of-mushroom-cultivation-in-greenhouses.

the IoT sensors situated within the mushroom bed.[381] The information is conveyed to the artificial intelligence model, which identifies the need for increased ventilation.[382] By stimulating actuators to activate fans embedded in the bed, the AI model regulates and maintains optimal CO2 levels.[383]

The goal is to develop a closed-loop IoT system that continuously collects real-time data from the mushroom cultivation environment using IoT sensors.[384] This data is processed by AI models to automatically optimize and regulate key growth conditions such as temperature, humidity, and substrate moisture.[385] A mobile or web application allows users to remotely monitor, receive alerts, and adjust environmental parameters as needed.[386] Additionally, the system ensures full traceability by tracking the origin and usage of substrates and supplements to maintain quality and safety standards.[387]

Mushroom Culinary Demand and Consumer Market

In Singapore's grocery stores, you can find a wide range of mushrooms to suit different tastes and Asian cooking styles.[388] White button mushrooms are a common type that can be used in a lot of different ways, from salads

---

[381] *Ibid.*

[382] *Ibid.*

[383] *Ibid.*

[384] Biotechnology, S. O. C. C. E. A. (2023). Internet of things (IoT)-based environmental monitoring and control system for home-based mushroom cultivation. NTU Singapore. https://dr.ntu.edu.sg/handle/10356/168857.

[385] Rohit Sharma, Sachin S. Kamble, Angappa Gunasekaran, Vikas Kumar, Anil Kumar, A systematic literature review on machine learning applications for sustainable agriculture supply chain performance, *Computers & Operations Research*, https://doi.org/10.1016/j.cor.2020.104926.

[386] Shanto, Shakib, Rahman, Musfiqur, Oasik, Jaber Md and Hossain, Hamim. (2023). Smart greenhouse monitoring system using Blynk IoT app. *Journal of Engineering Research and Reports*, 25. 94–107. 10.9734/JERR/2023/v25i2883.

[387] Tracextech and Tracextech. (2024a, September 29). 5-Step Guide to Traceability in Food Industry. Blockchain for Food Safety, Traceability and Supplychain Transparency. https://tracextech.com/traceability-in-food-industry.

[388] Fresh mushrooms | Shiitake, Swiss brown, oyster mushroom, black fungus. (n.d.). FairPrice. https://www.fairprice.com.sg/category/mushrooms.

to sauces.[389] Shiitake mushrooms are often used in some Asian cuisines, but especially in Japan, with a rich umami flavor.[390] Oyster mushrooms, with their delicate texture, are often used in stir-fries and stews, while the meaty portobello mushrooms serve as a great meat substitute in burgers and grilled dishes.[391] Enoki mushrooms are often added to soups, salads, and hot pots because they have a crunchy taste.[392]

Additionally, wood ear, white and black fungus, prized for their crunchy texture, are often found in Chinese soups and stir-fries.[393] Major supermarkets in Singapore, such as NTUC FairPrice, Cold Storage, and Sheng Siong, as well as a variety of wet markets and specialty organic stores, are all places where you may find these mushrooms.[394]

Industrial and Economic Potential

The mushroom industry shows great potential due to its wide-ranging applications, such as sustainable materials, food, medicine, and environmental remediation.[395] Notable companies like Adidas and Kering have forged partnerships with major players in the apparel sector to develop alternatives to leather.[396] Packaging material manufacturers are

---

[389] South Dakota Department of Health. (n.d.). *Health benefits and culinary uses of mushrooms.* https://healthysd.gov/category/nutrition/health-benefits-culinary-uses-of-mushrooms.

[390] Dunmore, G. (2022, September 29). Morning with Shiitake. The Japanese Pantry. https://thejapanesepantry.com/blogs/stories/morning-with-shiitake.

[391] Wysocarski, E. (2024b, September 23). Can mushrooms replace meat? 28 recipes that say yes. Olives for Dinner. https://olivesfordinner.com/mushroom-meat/.

[392] Fang, T. and Fang, T. (n.d.). Enoki is Just One of Many Japanese Mushrooms to Try! – Sakuraco. Sakuraco | Japanese Snacks & Candy Subscription Box. https://sakura.co/blog/enoki-is-just-one-of-many-japanese-mushrooms-to-try.

[393] Fan, S. and Fan, S. (2020, September 5). Salad of black and white wood ear mushrooms with cucumber | Soy, Rice, Fire. Soy, Rice, Fire | Inspiration From Classic Chinese Recipes and Techniques. https://soyricefire.com/2018/07/21/salad-of-black-and-white-wood-ear-mushrooms-with-cucumber.

[394] The 37 biggest supermarkets in Singapore in 2025. (n.d.). https://www.gourmetpro.co/blog/biggest-supermarkets-singapore.

[395] Ayodeji Amobonye, Japareng Lalung, Mukesh Kumar Awasthi, Santhosh Pillai, Fungal mycelium as leather alternative: A sustainable biogenic material for the fashion industry. *Sustainable Materials and Technologies,* https://doi.org/10.1016/j.susmat.2023.e00724.

[396] Okuda, Y. (2022). Sustainability perspectives for future continuity of mushroom production: The bright and dark sides. *Frontiers in Sustainable Food Systems,* 6. https://doi.org/10.3389/fsufs.2022.1026508.

collaborating with IKEA for strategic reasons, primarily its commitment to sustainability.[397] By partnering with these experts, IKEA aims to innovate eco-friendly packaging solutions that use renewable and recyclable materials.[398] Both of these applications of fungus-derived materials have already been put into practical use.[399]

While these products are not necessarily spent mushroom substrate or SMS-based, SMS is likely to be adopted as a cheaper material for them soon.[400] Thus, mushroom production has the potential to become a source of specific materials and food and to wipe out its "dark" side in ways that exceed their ecological role as decomposers in the natural ecosystem.[401]

Sustainability and Circular Practices

A smaller carbon footprint is also achieved through local farming, particularly through the utilization of recycled resources such as coffee grounds.[402] Utilizing coffee grounds as a substrate can be of great benefit to mushroom businesses since it can help to reduce the production costs by providing access to a waste material that is either inexpensive or free from coffee shops.[403] By recycling organic waste, this not only reduces costs but also contributes to the promotion of environmentally responsible activities.[404]

The cultivation of mushrooms in the country can foster economic growth by creating employment possibilities in agriculture and related areas.[405] A thriving industry requires a workforce with the right skills and talents, partnering with institutes of higher learning and

---

[397] *Ibid.*

[398] *Ibid.*

[399] *Ibid.*

[400] *Ibid.*

[401] *Ibid.*

[402] Sayner, A. (2023, December 11). Growing Mushrooms In Coffee Grounds. GroCycle. https://grocycle.com/growing-mushrooms-in-coffee-grounds/.

[403] *Ibid.*

[404] *Ibid.*

[405] FAO Knowledge Repository. (n.d.). https://openknowledge.fao.org/handle/20.500.14283/i0522e.

the industry to build a pipeline of talent for the sector and upskill the existing workforce.[406]

Sectoral Specialization in Mushroom Production

The three sectors of the specialty mushroom involve spawn production, block production, fruiting, and sales, each fulfilling specific functions in the production process.[407] Each of these three sectors can require completely different infrastructure, skill levels, and contacts to operate successfully.[408] As the small-scale specialty mushroom industry grows, there continues to be more specialization and focus for individual businesses.[409]

In commercial production of mushrooms, cultivation consists of six sequential steps that, while somewhat arbitrarily defined, collectively form a structured production system.[410] Step 1 (Phase I) involves compost preparation, Step 2 (Phase II) focuses on compost finishing, followed by Step 3, spawning, Step 4, casing, Step 5, pinning, and Step 6, cropping.[411] The industrial production of cultivated mushrooms, including Agaricus bisporus, Lentinula edodes (shiitake), and various Pleurotus species (oyster mushrooms), is traditionally based on substrates derived from the composting of organic waste, thereby distinguishing mushrooms as the sole agricultural crop that systematically utilizes and degrades solid waste materials.[412] In natural ecosystems, fungi perform an indispensable role in the decomposition of organic matter, the enhancement of soil organic

---

[406] "A sustainable food system for Singapore and beyond", published 11 November 2022, https://www.sfa.gov.sg/food-for-thought/article/detail/a-sustainable-food-system-for-singapore-and-beyond.

[407] Gabriel, S. (2020, April 6). The three sectors of the specialty mushroom – Cornell Small Farms. Cornell Small Farms. https://smallfarms.cornell.edu/2020/04/the-three-sectors-of-the-specialty-mushroom/.

[408] *Ibid.*

[409] *Ibid.*

[410] Daniel J. Royse. (2010). *Six steps to mushroom farming.* Pennsylvania State University. https://www.mushroomcompany.com/resources/agaricus/SixSteps.pdf.

[411] *Ibid.*

[412] Turners, C. (2022, June 24). How to produce commercial mushroom compost for mushroom farms? Affordable Compost Turner for Sale. https://compost-turner.net/composting-technologies/commercial-mushroom-compost-production-process.html.

content, and the recycling of minerals essential for sustaining soil fertility.[413]

An insightful article that delves into the critical environmental factors influencing mushroom cultivation is "7 Factors Affecting Mushroom Cultivation" by Atlas Scientific.[414] This piece outlines how variables such as temperature, pH, light, humidity, carbon dioxide (CO_2), moisture, and oxygen significantly impact the growth and yield of mushrooms.[415] However, the differential management of mushrooms according to growth stage, day and night cycles, production cycles, and seasonal variations is even more critical for achieving healthy and consistent mushroom production.[416]

Air Quality

Maintaining ambient air quality is vital as contaminants can adversely affect the cultivation environment.[417] Particulate matter sensors are employed within controlled environments; these sensors detect minute airborne hazards such as dust, smoke, and ultrafine particles.[418] They work by emitting light into the air with a laser diode and measuring the scattered light with a photodiode.[419] This data is processed by a microcontroller to determine mass concentration, accurately identifying particles categorized as PM 1.0, PM 2.5, and PM 10.[420]

These sensors are designed for easy integration into various applications, including indoor and outdoor air quality monitoring, HVAC system

---

[413] *Ibid.*

[414] Pomelo. (2023b, April 13). 7 Factors Affecting Mushroom Cultivation | Atlas Scientific. Atlas Scientific. https://atlas-scientific.com/blog/factors-affecting-mushroom-cultivation.

[415] *Ibid.*

[416] MIRAE C&H. (2022, July 6). Mushroom Cultivation Solution | MIRAE C&H. Mushroom Cultivation System, Hydroponic System, Greenhouse System | MIRAE C&H -. http://miraecnh.com/en/mushroom/.

[417] P. I. E. (2021, April 7). Panasonic particulate matter sensor SN-GCJA5 – air quality and dust sensor. YouTube. https://www.youtube.com/watch?v=6D5Zz3AckPc.

[418] *Ibid.*

[419] *Ibid.*

[420] *Ibid.*

regulation, and IoT devices.[421] They are robust against dust, have a long lifespan, offer rapid responsiveness, and provide precise and reliable results.[422] The device offers precise PM1, PM2.5, and PM10 detection using a laser diode and microprocessor-based analysis, featuring auto-calibration, optimized airflow to reduce dust buildup, and mass-density data output via I2C and UART, making it ideal for HVAC, air quality monitoring, smart homes, and IoT applications.[423]

Sterilization and Lab Protocols

The cultivation process begins in the media preparation and sterilization room, where substrates are cleaned to eliminate pathogens.[424] Here, substrates are carefully prepared and sanitized to ensure they are free from contaminants, laying the groundwork for successful mushroom cultivation by encouraging the growth of mycelial organisms.[425]

The basic principle of steam sterilization, as accomplished in an autoclave, is to expose each item to direct steam contact at the required temperature and pressure for the specified time.[426] Autoclaves are also known as steam sterilizers and are typically used for healthcare or industrial applications.[427] An autoclave is a machine that uses steam under pressure to kill harmful bacteria, viruses, fungi, and spores on items that are placed inside a pressure vessel.[428] The items are heated to a specific sterilization temperature for a set duration, during which the steam's

---

[421] Automated Internet of Things system for monitoring indoor air quality. (2024). In CEUR Workshop Proceedings (p. 97) [Journal-article]. https://ceur-ws.org/Vol-3666/paper15.pdf.

[422] *Ibid.*

[423] Gough Lui. (2022, March 23). *The Panasonic SN-GCJA5 particulate matter sensor.* element14 Community. https://community.element14.com/products/roadtest/rv/roadtest_reviews/1570/the_panasonic_sn-gcj.

[424] Beyer, D. M. *Op.cit.*

[425] *Ibid.*

[426] Steam sterilization. (2023, November 28). Infection Control. https://www.cdc.gov/infection-control/hcp/disinfection-sterilization/steam-sterilization.html.

[427] Autoclave Machine: Uses, guidelines & Cost | Knowledge Center. (n.d.). https://www.steris.com/healthcare/knowledge-center/sterile-processing/everything-about-autoclaves.

[428] *Ibid.*

moisture efficiently transfers heat to destroy the protein structures of bacteria and spores.[429]

A mushroom autoclave operates by sealing substrate bags inside a chamber, where steam displaces air and gradually increases temperature and pressure during the purge phase.[430] Once the desired conditions are reached, the sterilization phase maintains heat for a set duration before the exhaust phase releases pressure, and an optional vacuum cycle ensures complete drying.[431]

Innovation in Smart Agriculture and Sensor Integration

Greiner Assistec and Accensors, a unit of innoME GmbH, are collaborating on a project to develop printed sensor systems, with their first major success being an intelligent petri dish prototype.[432] The dish incorporates sensors printed on a film that is thermoformed into a petri dish shape, allowing continuous monitoring of pH and temperature via a scanner and app.[433] The technology promises cost-effective, high-volume production while maintaining sensor functionality through the thermoforming process.[434] This innovation highlights the potential for transforming plastic parts with printed electronics, supporting the establishment of a new center for smart plastics solutions in Austria, and opening future applications in biotech, smart manufacturing, and agriculture.[435]

---

[429] *Ibid.*

[430] Mechler, S. (2024, October 15). Autoclaves for mushroom cultivation. Consolidated Sterilizer Systems. https://consteril.com/mushroom-autoclave/.

[431] *Ibid.*

[432] Greiner Packaging International GmbH. (n.d.). Greiner Assistec working in strong partnership to develop. https://www.greiner-assistec.com/en/Press/Greiner-Assistec-working-in-strong-partnership-to-develop-plastic-parts-with-printed-sensor-systems_s_294217.

[433] Gul, J. Z., Khan, M., Rehman, M. M., Mohy Ud Din, Z., and Kim, W. Y. (2023). Preparation and performance analysis of 3D thermoformed fluidic polymer temperature sensors for aquatic and terrestrial applications. *Sensors*, 23(20), 8506. https://doi.org/10.3390/s23208506.

[434] *Ibid.*

[435] New Technology Leads to Smart Petri Dish Development. (n.d.). Labmate Online. https://www.labmate-online.com/news/laboratory-products/3/greiner-assistec/new-

Liquid Cultures and Laboratory Techniques

The process of propagating fungi through liquid culture involves creating a nutrient-rich solution in a sterile environment to support mycelium growth.[436] Growers propagate mycelium by inoculating a nutrient-rich medium with mushroom tissue or spores, resulting in the formation of either a petri dish culture or a liquid culture.[437] Initially, a Liquid Culture Medium (LCM) is prepared using ingredients like malt extract and yeast extract, and sterilized to eliminate contaminants.[438] Mushroom tissue or spores are introduced into the medium, which is then incubated to allow mycelium growth.[439] After incubation, the mycelium-infused culture is used to inoculate larger substrates in grow bags.[440] These bags are moved to an incubation room where the mycelium spreads through the substrate.[441] In the fruiting room, optimized conditions such as temperature and humidity stimulate mushroom growth.[442]

In highly controlled laminar flow rooms, biosafety cabinets ensure sterile conditions for handling mushroom cultures.[443] These cabinets maintain a clean environment, crucial for sterile transfers and minimizing contamination risks.[444] This meticulous approach supports high-quality

---

technology-leads-to-smart-petri-dish-development/55139.

[436] Sayner, A. (2024, October 16). Mushroom Inoculation: Spawn, Substrate, Logs & Beds. GroCycle. https://grocycle.com/mushroom-inoculation/.

[437] *Ibid.*

[438] Sayner, A. (2024, May 8). How to Make Liquid Culture for Mushrooms: The Complete Guide. GroCycle. https://grocycle.com/how-to-make-liquid-culture/.

[439] *Ibid.*

[440] Kenneth Kanayo Alaneme, Justus Uchenna Anaele, Tolulope Moyosore Oke, Sodiq Abiodun Kareem, Michael Adediran, Oluwadamilola Abigael Ajibuwa, Yvonne Onyinye Anabaranze, Mycelium based composites: A review of their bio-fabrication procedures, material properties and potential for green building and construction applications. *Alexandria Engineering Journal*, Volume 83, 2023, Pages 234–250, ISSN 1110-0168, https://doi.org/10.1016/j.aej.2023.10.012.

[441] *Ibid.*

[442] *Ibid.*

[443] Laminar Flow Hoods for Mushroom Cultivation, Horticulture, Mycology, Micropropagation. (2022, September 20). Terra Universal. Retrieved March 23, 2025, from https://www.terrauniversal.com/blog/laminar-flow-hoods-for-mushrooms-mycology-mold-yeast-fungus.

[444] *Ibid.*

mushroom production, safeguarding both worker safety and crop integrity.[445]

For optimal mycelium development, it is essential to provide high nutrient availability and a moist environment. [446] All nutrient-rich food-based substrates like rye grain, popcorn, brown rice, and wheatberries must be sterilized, as they readily attract various fungi and molds.[447]

Liquid culture media (broth) and petri dish agar are both used to cultivate microorganisms, but liquid media support uniform growth, while agar plates are used for isolating colonies and studying microbial growth patterns.[448] Agar media is commonly employed due to its gel-like consistency and nutrient content, which is ideal for controlled cultivation and strain isolation.[449] Agar is a gelatinous substance derived from the cell walls of certain red algae species, most notably "ogonori" and "tengusa."[450] Agar is a complex carbohydrate derived from seaweed.[451] In mycology, a petri dish creates a controlled environment to observe and study the growth of fungi, whether for research, cloning, strain selection, or simply learning.[452] The choice of growing medium is crucial in mycology, as it directly affects the growth and health of mushrooms. Growing media come in the form of powders to mix with water and are enriched with various nutrients to encourage the growth of fungi.[453]

---

[445] *Ibid.*

[446] Webb, E. and It, U. F. (2025, February 10). Ultimate Guide to mushroom Substrates. Urban Farm-It. https://urban-farm-it.com/blogs/mushroom-cultivation/guide-to-mushroom-substrates.

[447] Webb, E. and It, U. F. (2025, February 10). Ultimate Guide to mushroom Substrates. Urban Farm-It. https://urban-farm-it.com/blogs/mushroom-cultivation/guide-to-mushroom-substrates.

[448] Fatima, A. (2024, May 25). Microbial Culture Media: types, examples, uses. Microbe Notes. https://microbenotes.com/types-of-culture-media/.

[449] FloCube. (2023, September 22). What is Agar used for in Mushroom Cultivation? – FloCube – Medium. Medium. https://medium.com/@flocube/what-is-agar-used-for-in-mushroom-cultivation-9977dc4d26e4.

[450] *Ibid.*

[451] *Ibid.*

[452] How to make petri dishes for mycological cultivation: Guide for beginners. (2024, September 22). La Mycosphère. https://lamycosphere.com/en-fr/blogs/the-future-is-fungi/how-to-make-petri-dishes-for-mycological-culture-guide-for-beginners.

[453] *Ibid.*

Fungi as a Bioremediation and the New Material

Fungi play a vital role in ecosystems by breaking down dead plants and animals through powerful enzymes, recycling essential nutrients like carbon, nitrogen, and phosphorus back into the soil for other organisms to use.[454] In addition to decomposition, fungi help regulate disease by removing animal carcasses from the landscape and maintain ecosystem balance by breaking down fallen trees and plant debris.[455] While fungi are essential for ecosystem health, the specific contributions of fungal diversity to these functions remain not fully understood.[456] Beyond decomposition, fungi also form symbiotic relationships with plants and contribute to carbon sequestration, making them indispensable to life on Earth.[457] As mycologists uncover more about the remarkable biochemical properties of fungi, growing evidence supports Paul Stamets' assertion that "mycelium is the neurological network of nature."[458]

The root-associated microbiomes, which include fungi, bacteria, archaea, protists, and viruses, engage in a biochemical exchange that supports plant health and vitality.[459] Notably, mycorrhizal fungi stand out among their groups, establish themselves within the root systems of their botanical hosts, and facilitate the enhanced absorption of essential nutrients, notably phosphorus and nitrogen.[460] Fungi, particularly mycelium, have garnered significant attention for their potential to address ecological issues, making them crucial environmental guardians.[461] Mycelium, the vegetative part of fungi, excels in bioremediation due to its ability

---

[454] Horlick, C. (2023, July 5). Fungi: The Hidden Heroes of Ecosystems. Earth.Org. https://earth.org/fungi-the-hidden-heroes-of-ecosystems/.

[455] *Ibid.*

[456] *Ibid.*

[457] Bonfante, P. and Venice, F. (2020). Mucoromycota: going to the roots of plant-interacting fungi. *Fungal Biology Reviews*, 34(2), 100–113. https://doi.org/10.1016/j.fbr.2019.12.003.

[458] Horlick, C. (2023, July 5). Fungi: the hidden heroes of ecosystems. Earth.Org. https://earth.org/fungi-the-hidden-heroes-of-ecosystems.

[459] Shi, J., Wang, X., and Wang, E. (2023). Mycorrhizal symbiosis in plant growth and stress adaptation: From genes to ecosystems. *Annual Review of Plant Biology*, 74(1), 569–607. https://doi.org/10.1146/annurev-arplant-061722-090342.

[460] *Ibid.*

[461] *Ibid.*

to metabolize pollutants and purify contaminated substances.[462] This capability makes it invaluable for ecological restoration efforts aimed at mitigating environmental degradation.[463]

Increasing urbanization and industrialization are worsening global environmental pollution, highlighting the need for sustainable solutions, and fungi offer a promising, eco-friendly, and cost-effective approach to waste management.[464] Researchers are exploring innovative methods to utilize specific fungal strains to break down and metabolize waste materials such as plastic gloves and polypropylene face masks, which proliferated during the COVID-19 pandemic.[465]

Wildfires in the San Francisco Bay Area are becoming more frequent and intense due to climate change and human activities, prompting ecologists to use oyster mushrooms to restore scorched soil by breaking down toxic ash into bioavailable compounds.[466] Fungi show great promise in remediating polluted areas by breaking down toxic materials, but this science is still in its early stages, requiring further research and passionate innovators to harness fungi's potential for environmental restoration.[467]

A biotech company in upstate New York uses fast-growing mycelium to create biodegradable packaging as an eco-friendly alternative to Styrofoam.[468] By selecting suitable fungal species and feeding mycelium with agricultural waste, it can be grown into molds where it naturally binds together, forming solid structures that can be dried and used as biodegradable alternatives to Styrofoam.[469] However, this field is still

---

[462] Yuvaraj Dinakarkumar, Gnanasekaran Ramakrishnan, Koteswara Reddy Gujjula, Vishali Vasu, Priyadharishini Balamurugan, Gayathri Murali, Fungal bioremediation: An overview of the mechanisms, applications and future perspectives, *Environmental Chemistry and Ecotoxicology*, https://doi.org/10.1016/j.enceco.2024.07.002.

[463] *Ibid.*

[464] McKenna, J. (2023, December 12). Using fungi for waste management. MDPI Blog. https://blog.mdpi.com/2023/12/12/fungi-for-waste-management/.

[465] E. (2022, February 3). Mycoremediation: How Fungi Can Repair Our Land. Office of Sustainability – Student Blog. https://usfblogs.usfca.edu/sustainability/2022/02/03/mycoremediation-how-fungi-can-repair-our-land/.

[466] *Ibid.*

[467] *Ibid.*

[468] B. I. (2021, April 11). How Mushrooms Are Turned Into Bacon And Styrofoam | World Wide Waste. YouTube. https://www.youtube.com/watch?v=uznXI8wrdag.

[469] *Ibid.*

developing, and more research and dedicated efforts are needed to fully harness fungi's natural ability to heal the environment.[470]

Because of their natural ability to break down contaminants, fungi are well-suited for cleaning up polluted areas, removing heavy metals, reducing plastic pollution, and dealing with oil spills.[471] The mycelium releases enzymes to catalyze the decomposition of poisons into inactive molecules, effectively outwardly digesting contaminants before they are absorbed.[472] These enzymes function as biological cleavers, specifically targeting cellulose, lignin, and chitin, which are present in the cell walls of plants and the exoskeletons of insects.[473]

Fungal species have evolved to flourish in diverse environments, spanning from extremely cold arctic regions to extremely hot deserts.[474] Fungi can break down organic substances in environments with low levels of oxygen; this flexibility guarantees that decomposition occurs effectively worldwide.[475] Fungal enzymes are classified into seven major classes, including transferases, oxidoreductases, lyases, hydrolases, ligases, isomerases, and translocases.[476] They exhibit both intracellular and extracellular activity, playing a crucial role in decomposition, bioconversion, and protection against hazardous compounds, with commercially significant compounds being extensively studied for xenobiotic degradation and industrial wastewater treatment.[477] However, their structure and

---

[470] *Ibid.*

[471] From Ecuador To New Zealand: How Fungi Can Help Clean Oil Spills | The Momentum. (n.d.). https://www.themomentum.com/articles/from-ecuador-to-new-zealand-how-fungi-can-help-clean-oil-spills.

[472] *Ibid.*

[473] Jayasekara, S. and Ratnayake, R. (2019). Microbial cellulases: An overview and applications. IntechOpen. https://www.intechopen.com/chapters/66517.

[474] Cene Gostinčar, Polona Zalar, Nina Gunde-Cimerman, No need for speed: Slow development of fungi in extreme environments. *Fungal Biology Reviews*, https://doi.org/10.1016/j.fbr.2021.11.002.

[475] Chapter 1, The decomposition Process – Earth-Kind® Landscaping Earth-Kind® Landscaping. (n.d.). https://aggie-horticulture.tamu.edu/earthkind/landscape/dont-bag-it/chapter-1-the-decomposition-process.

[476] El-Gendi, H., Saleh, A. K., Badierah, R., Redwan, E. M., El-Maradny, Y. A., and El- Fakharany, E. M. (2021). A comprehensive insight into fungal enzymes: Structure, classification, and their role in mankind's challenges. *Journal of fungi (Basel, Switzerland)*, 8(1), 23. https://doi.org/10.3390/jof8010023.

[477] *Ibid.*

diverse biotechnological applications in industry, medicine, bioremediation, and environmental sustainability are critically reviewed.[478]

Fungi's extensive network of filamentous structures permeates the soil, excreting potent digestive enzymes that enable the breakdown of even the most resilient materials.[479] Mycoremediation offers a promising method for environmental remediation, resulting in soil that is not only free of contaminants but also fertile and enhanced with carbon.[480]

While science supports the efficacy of mycoremediation, there are still barriers to widespread deployment.[481] As researchers continue to delve into the intricacies of mycoremediation and refine its methodologies, it holds immense promise for the widespread adoption of fungi-based remediation strategies.[482] Mycoremediation, particularly through the use of native fungi, is one of many tools for community restoration projects aimed at regenerating areas hit hard by human-made hazards, where erosion, decay, disaster, pollution, or mismanagement have caused the ecosystems to falter.[483] This method utilizes the distinct abilities of fungi to break down various pollutants, such as heavy metals, agricultural waste, and pharmaceuticals.[484]

There is a need for studies focusing on optimizing the conditions inside the bioreactors for the effective mycoremediation process.[485] The indigenous fungi growing on polluted sites should be considered for further studies as they are adapted to the high concentration of various

---

[478] *Ibid.*

[479] Dykes, J. (2022, May 20). Mycoremediation: The under-utilised art of fungi clean-ups. Geographical. https://geographical.co.uk/science-environment/mycoremediation-the-art-of-fungi-clean-ups.

[480] *Ibid.*

[481] From Ecuador To New Zealand: How Fungi Can Help Clean Oil Spills | The Momentum. *Op.cit.*

[482] *Ibid.*

[483] Hance, J. (2021, September 15). Mycoremediation brings the fungi to waste disposal and ecosystem restoration. Mongabay Environmental News. https://news.mongabay.com/2021/09/mycoremediation-brings-the-fungi-to-waste-disposal-and-ecosystem-restoration/.

[484] Yuvaraj Dinakarkumar, Gnanasekaran Ramakrishnan, Koteswara Reddy Gujjula, Vishali Vasu, Priyadharishini Balamurugan, Gayathri Murali, *Op.cit.*

[485] Nahid Akhtar, M. Amin-ul Mannan, Mycoremediation: Expunging environmental pollutants. *Biotechnology Reports*, https://doi.org/10.1016/j.btre.2020.e00452.

pollutants in harsh environmental conditions.[486] Nevertheless, there is encouraging evidence to support that mycoremediation can expunge the environmental pollutants and make this planet a safe habitat.[487]

These products demonstrate a shift towards finding solutions that balance human innovation with the need to protect and care for the environment.[488] Using mycelium-based material as an alternative to polystyrene and plastic packaging was started in 2013 by a company called Ecovative Design.[489] Companies in the United States like MycoWorks and Ecovative are leading innovations with products like *reishi* leather and mycelium packaging.[490] These materials offer benefits such as fire resistance, thermal and acoustic insulation, and minimal resource use.[491] However, scaling up production poses challenges, including energy requirements for growth and drying processes.[492]

The market for these materials is expected to grow significantly, with increasing investments and consumer interest.[493] In Asia, Mycotech Lab (MYCL), a bio-based material startup from Indonesia, utilizes mushroom mycelium to produce environmentally friendly substitutes for leather and construction materials.[494] In order to widen the use of its method, MYCL has collaborated with research hubs, such as the Singapore-ETH Centre in 2016, to develop new versions of its biomaterials.[495] The company converts mycelium and agricultural waste into long-lasting, environmentally

---

[486] *Ibid.*

[487] *Ibid.*

[488] Alemu, D., Tafesse, M., and Mondal, A. K. (2022b). Mycelium-Based Composite: The future sustainable biomaterial. International Journal of Biomaterials, 2022, 1–12. https://doi.org/10.1155/2022/8401528.

[489] *Ibid.*

[490] Navasiolava, L. R. T. (2024, October 1). Mushrooms are now becoming leather, packaging, bacon and more. Washington Post. https://www.washingtonpost.com/business/interactive/2024/mushroom-mycelium-sustainability-leather-shoes/.

[491] Kenneth Kanayo Alaneme, Justus Uchenna Anaele, Tolulope Moyosore Oke, Sodiq Abiodun Kareem, Michael Adediran, Oluwadamilola Abigael Ajibuwa, Yvonne Onyinye Anabaranze, *Op.cit.*

[492] *Ibid.*

[493] Hernandez, A. (2024, January 14). Mycomaterials, Mycelium-Based Materials and the Future of Fungi — Myceliumatters. Myceliumatters. https://myceliummatters.org/myconews/mycomaterials-myceliumbasedmaterials.

[494] *Ibid.*

[495] *Ibid.*

beneficial products named Mylea, a material that resembles leather.[496] The company combines quick prototyping techniques and engages in collaborations with commercial partners to create durable products such as fire boards and fashionable accessories.[497] According to MYCL CEO Adi Reza Nugroho, the company's alternative material production process demonstrates substantially lower environmental demands than traditional leather tanning, including major decreases in carbon emissions, water use, and overall energy requirements.[498] The company has also created innovative mycelium tiles for use as internal wall cladding in an experimental housing project set on the Indonesian island of Batam.[499] Reza expressed pride that the company's product had been selected for a government-backed initiative focused on developing sustainable and affordable housing solutions for low-income communities in developing nations.[500]

Mycelium materials may help reduce carbon emissions by replacing petroleum-based materials such as plastics and synthetic foams commonly used in packaging, insulation, and consumer products.[501] Because fungi grow naturally on agricultural waste substrates such as sawdust, straw, and corn husks, the production process requires significantly less energy compared to conventional manufacturing methods that rely on fossil fuel-derived raw materials.[502] In addition, the biological growth process allows mycelium to bind organic particles together while capturing and storing carbon within the material structure.[503] This natural fabrication process minimizes industrial processing and chemical inputs. As a result, mycelium-based materials offer a promising pathway toward

---

[496] *Ibid.*

[497] *Ibid.*

[498] Why MYCL's mushroom mycelium could be the basis of your next bag – or house. (2023, January 3). SG Innovate. Retrieved June 14, 2024, from https://www.sginnovate. com/blog/why-mycls-mushroom-mycelium-could-be-basis-your-next-bag-or-house.

[499] *Ibid.*

[500] *Ibid.*

[501] Alaneme, K. K., Anaele, J. U., Oke, T. M., Kareem, S. A., Adediran, M., Ajibuwa, O. A., and Anabaranze, Y. O. (2023). Mycelium-based composites: A review of their bio-fabrication procedures, material properties and potential for green building and construction applications. *Alexandria Engineering Journal, 83,* 234–250. https://doi. org/10.1016/j.aej.2023.10.012.

[502] *Ibid.*

[503] *Ibid.*

low-carbon manufacturing and more sustainable production systems within emerging circular bioeconomy frameworks.[504]

The transition toward a circular economy is increasingly supported by emerging bio-innovation strategies that transform organic waste into valuable resources.[505] Urban environments generate significant quantities of agricultural and food residues, which can serve as substrates for biological production systems such as mushroom cultivation and mycelium-based biomaterials.[506] By converting organic waste streams into useful products, cities can reduce landfill dependency while supporting sustainable manufacturing.[507] Integrating biotechnology with urban resource management also creates opportunities for decentralized production systems, where biological materials are cultivated locally using controlled environments. Such approaches contribute to resource efficiency, lower environmental impact, and more resilient urban sustainability frameworks.[508] Mycelium-based materials offer numerous advantages; they're low-cost, biodegradable, fire-resistant, and serve as both thermal and acoustic insulators.[509] These sustainable innovations have the potential to replace conventional materials used in packaging, fashion, and single-use products.[510] While full integration into mainstream society is still underway, companies like Bolt Threads and MycoWorks are leading the shift, collaborating with major brands such as Stella McCartney, Lululemon, and Adidas to introduce biodegradable products made from Mylo and Reishi Leather.[511] Mycomaterials are created by growing mycelium, the root-like structure of fungi, on agricultural waste, then shaping, drying, and processing it into durable, biodegradable products, offering a sustainable and eco-friendly alternative to conventional materials like leather and plastic.[512]

In Singapore, achieving zero waste remains challenging due to limited public awareness of recycling practices, with the National Environment Agency (NEA) reporting that 40% of blue bin contents are contaminated

---

[504] *Ibid.*

[505] Grimm, D. and Wösten, H. A. B. (2018). Mushroom cultivation in the circular economy. *Applied Microbiology and Biotechnology*, 102(18), 7795–7803. https://doi.org/10.1007/s00253-018-9226-8.

[506] *Ibid.*

[507] *Ibid.*

[508] *Ibid.*

[509] Hernandez, A. (2024, January 14). *Op.cit.*

[510] *Ibid.*

[511] *Ibid.*

[512] *Ibid.*

by non-recyclables such as food waste.[513] Bioremediation plays a vital role in advancing circular economy goals by addressing pollution and supporting resource recovery.[514] Using living organisms like fungi, bacteria, and plants, this method breaks down pollutants and restores contaminated environments, enhancing both sustainability and biodiversity.[515] Additionally, bioremediation facilitates resource recovery, reduces the need for raw material extraction, and promotes circularity by minimizing overall waste generation.[516]

Mycelium in a Circular Economy

In 2020, the City of Amsterdam pledged to transition to a circular economy, setting a goal to eliminate waste and fully recycle all materials by 2050.[517] The city council outlined clear goals, as the study indicated that the transition would lead to a cleaner environment, increased employment, and a stronger economy.[518] These include achieving a full circular economy by 2050, ensuring 65% of household waste is reusable by 2025, and reducing the use of primary raw materials by 50% by 2030 compared to 2019.[519]

Mycelium-based materials represent a significant stride towards embracing the Circular Economy model, where resources are used more efficiently, and waste is minimized through continual use and recycling.[520] By using the regenerative power of nature, these bio-based products offer

---

[513] Waste management. *Op.cit.*

[514] Bioremediation as an effective tool for a clean environment. (2024, September 3). *Op.cit.*

[515] *Ibid.*

[516] *Ibid.*

[517] The Earthshot Prize. (2024, August 27). City of Amsterdam, Circular Economy – the Earthshot Prize. https://earthshotprize.org/winners-finalists/city-of-amsterdam-circular-economy.

[518] Kuttan, S. C. (2021, March 22). Commentary: Taking Singapore's green ambitions to new heights with a circular economy. CNA. https://www.channelnewsasia.com/commentary/singapore-circular-economy-sustainability-community-climate-315981.

[519] *Ibid.*

[520] Digafe Alemu, Mesfin Tafesse, and Ajoy Kanti Mondal. (2022, March 12). Mycelium-Based Composite: The Future Sustainable Biomaterial. National Library of Medicine. Retrieved June 13, 2024, from https://www.ncbi.nlm.nih.gov/pmc/articles/PMC8934219/.

a sustainable alternative to conventional materials across various indus-tries.[521] In a circular economy, resources are used in a way that prevents waste, materials are reused, repaired, or recycled continuously, and natu-ral systems are restored rather than depleted, creating a sustainable cycle that benefits both the environment and the economy.[522] Products and materials are kept in circulation through processes like maintenance, reuse, refurbishment, remanufacture, recycling, and composting.[523]

The circular economy tackles climate change and other global challenges, e.g., biodiversity loss, waste, and pollution, by decoupling economic activity from the consumption of finite resources.[524] At the core of the Circular Economy lies the principle of "closing the loop", ensuring that products and materials are reused, repurposed, or recycled at the end of their life cycle rather than being disposed of as waste.[525] A circular economy is a system designed to be regenerative by design, ensuring that products and materials are reused, recycled, or repurposed to extend their lifecycle and minimize waste.[526]

The role of mycelium-based materials in advancing sustainability within the circular economy is increasingly recognized, offering an eco-friendly alternative to conventional materials across sectors such as packaging, construction, and textiles.[527] In addition to providing advan-tages such as cost-effectiveness, a minimal environmental footprint, and recyclability, their intrinsic biodegradability allows seamless reintegration into natural ecosystems without ecological harm.[528] Mycomaterials pos-sessed beneficial properties, including fire resistance and superior thermal

---

[521] *Ibid.*

[522] Circular Economy Introduction. (n.d.). https://www.ellenmacarthurfoundation.org/topics/circular-economy-introduction/overview.

[523] *Ibid.*

[524] Puffpaff, M. (2023, September 5). The circular economy. National Energy Foundation. https://nef.org.uk/the-circular-economy/.

[525] Julian Kirchherr, Nan-Hua Nadja Yang, Frederik Schulze-Spüntrup, Maarten J. Heerink, and Kris Hartley, Conceptualizing the circular economy (Revisited): An analysis of 221 definitions, resources, conservation and recycling, https://doi.org/10.1016/j.resconrec.2023.107001.

[526] Circular economy introduction. *Op.cit.*

[527] Mitchell Jones, Andreas Mautner, Stefano Luenco, Alexander Bismarck, Sabu John, Engineered mycelium composite construction materials from fungal biorefineries: A criti-cal review. *Materials & Design*, https://doi.org/10.1016/j.matdes.2019.108397.

[528] *Ibid.*

and acoustic insulation, as mentioned.[529] Although the widespread integration of mycelium-based materials into mainstream industries remains a developing endeavor, numerous enterprises are already progressing toward more sustainable practices.[530] Current applications extend across diverse fields, including manufacturing and biomedical industries, where mycelium-based composites demonstrate significant potential for reducing environmental impact and promoting economic sustainability.[531]

Concluding Remarks

Mushrooms have played a significant role in human history, revered for their culinary, medicinal, and spiritual properties across diverse cultures. Contemporary research affirms their value as nutrient-dense, low-calorie foods rich in protein, fiber, vitamins, and bioactive compounds with potential applications in disease prevention and cognitive enhancement. Globally, gourmet and functional mushrooms such as shiitake, oyster, lion's mane, and *reishi* are increasingly sought after for their health benefits and culinary versatility. Simultaneously, sustainable mushroom cultivation has emerged as a viable solution to environmental and food security challenges. Compared to conventional livestock, mushrooms require significantly less land, water, and energy, positioning them as an environmentally responsible protein alternative.

In Singapore, the limited availability of arable land and dependence on food imports have accelerated the development of high-tech, space-efficient mushroom farming systems. Vertical farms employing artificial intelligence (AI), the Internet of Things (IoT), and climate-controlled environments have enabled year-round, high-yield production. Farms such as Golden Cap and Finc Bio-Tech exemplify the integration of automation, real-time data analytics, and circular economy principles, including the upcycling of agricultural waste into mushroom substrates.

In the future, mushrooms are expected to play an increasingly central role in Singapore's agri-food landscape. As consumer demand for healthy, locally sourced, and sustainable foods continues to rise, investment in

---

[529] Hernandez, A. (2024, January 14). *Op.cit.*

[530] *Ibid.*

[531] Emma Camilleri, Sumesh Narayan, Divnesh Lingam, Renald Blundell, Mycelium-based composites: An updated comprehensive overview. *Biotechnology Advances*, https://doi.org/10.1016/j.biotechadv.2025.108517.

smart farming infrastructure and biotech innovation is likely to expand. Advances in AI-driven precision farming, substrate optimization, and functional food development will further enhance the nutritional value, production efficiency, and market reach of mushroom products. Additionally, mycelium-based materials are poised to support Singapore's zero-waste and bioeconomy ambitions by offering eco-friendly alternatives in packaging, textiles, and construction. Thus, the future of mushrooms in Singapore lies not only in feeding the population but also in driving regenerative, circular, and technologically enabled urban agriculture.

Agricultural Challenges in Singapore

Singapore's "30 by 30" goal aims to meet 30% of the nation's nutritional needs through local food production by 2030.[532] Achieving this target requires substantial innovation and transformation within the agricultural sector.[533] Key initiatives include the redevelopment of Lim Chu Kang into a high-tech agricultural district, and the adoption of advanced methods such as stacked and underground farming, as well as investments in renewable energy for greenhouse operations.[534]

In addition to boosting food production, efforts have been made to diversify the types of food that can be grown locally.[535] Established in 2019, the Singapore Institute of Food and Biotechnology Innovation (SIFBI) under A*STAR focuses on researching alternative proteins and food ingredients.[536] Collaborating with Nurasa, a food-tech accelerator owned by Temasek Holdings, SIFBI has committed over $30 million to support the growth of food-tech startups.[537]

---

[532] Tan, J. (2024, April 1). Farmers face hurdles in push to innovate and boost Singapore's food security. The Straits Times. https://www.straitstimes.com/singapore/farmers-face-hurdles-in-push-to-innovate-and-boost-singapore-s-food-security.

[533] *Ibid.*

[534] *Ibid.*

[535] AgroSpectrum. (2024, June 7). Innovation, Prediction, and Flexibility: three pillars to ensure a resilient food supply chain for Singapore. TMX. Retrieved March 28, 2025, from https://tmxtransform.com/blog/innovation-prediction-and-flexibility-three-pillars-to-ensure-a-resilient-food-supply-chain-for-sinapore/.

[536] *Ibid.*

[537] *Ibid.*

Despite these advancements, transitioning to a high-tech food industry remains challenging.[538] Farmers face high operational costs, limited land availability, and competition for land use.[539] Some traditional farms have struggled to adopt new technologies and have been forced to shut down.[540]

Agri-tech consultant Lionel Wong, co-founder of Upgrown Farming, often advises against entering the vertical vegetable farming sector without substantial capital.[541] Drawing from over a decade of experience, he has seen many startups fail due to unsustainable business models.[542] Similarly, Mr. Kenny Eng, director of horticulture firm Nyee Phoe Group, emphasizes that technological innovation alone is insufficient.[543] He advocates for a holistic approach that integrates technology with social resilience and ecological balance.[544]

One notable case is I.F.F.I., whose parent company, TranZPlus Engineering, entered liquidation in November 2023.[545] Despite its 38,000 sq m facility in Tuas featuring AI-driven farming systems and advanced environmental and water treatment technologies, the farm ceased operations.[546] I.F.F.I. had been a recipient of the Singapore Food Agency's "30 by 30 Express" grant, receiving funding as part of a $39.4 million package awarded to nine farms during the COVID-19 pandemic.[547] Its closure highlights the operational and financial difficulties faced by high-tech agricultural ventures.[548]

---

[538] Loh, R. (2024c, October 21). With several farms closing or struggling to break even, what is the future for agriculture in Singapore? CNA. https://www.channelnewsasia.com/today/big-read/high-tech-low-returns-farming-4684566.

[539] *Ibid.*

[540] *Ibid.*

[541] Loh, R. (2024c, October 21). *Op.cit.*

[542] *Ibid.*

[543] GAadmin. (2020, February 24). The future of Singapore's food security. Gardenasia. https://www.gardenasia.com/2020/02/24/the-future-of-singapores-food-security/.

[544] *Ibid.*

[545] Tan, C. (2024, May 27). Mega indoor veggie farm I.F.F.I shuts down in latest blow to local food production. The Straits Times. https://www.straitstimes.com/singapore/mega-indoor-veggie-farm-iffi-shuts-down-in-latest-blow-to-local-food-production.

[546] *Ibid.*

[547] *Ibid.*

[548] *Ibid.*

Cost remains a significant barrier to consumer adoption of local produce.[549] Due to high expenses related to manpower, utilities, and land leases, locally grown crops are often more expensive than imported alternatives.[550] Melvin Tan, director of Blue Ocean Harvest, notes that production costs in Singapore can be three times higher than in neighboring Malaysia, making it difficult for local farms to remain competitive especially among price-sensitive consumers.[551]

Other projects have encountered delays. ISE Foods Holdings (IFH), Singapore's fourth egg farm, has faced setbacks due to rising construction costs and biosecurity implementation challenges.[552] Nevertheless, the farm is seen as a critical addition to Singapore's food security, with plans to produce 360 million eggs annually an increase from the 29% of domestic egg demand met in 2022.[553] IFH intends to operate an integrated supply chain, including hatcheries and parent farms in various locations.[554]

Meanwhile, VertiVegies, which once aimed to build one of Singapore's largest indoor farms, cancelled its plans due to partnership conflicts with SananBio, supply chain issues, and failure to meet the Singapore Food Agency's construction deadlines.[555] The company returned the land in April 2022 and is now reassessing its strategy to align with the "30 by 30" target.[556]

---

[549] Ali, N. H. M. (2023, June 11). The Big Read: Buy local but at what price? Costlier Singapore produce could hamper quest for greater food security. CNA. https://www.channelnewsasia.com/today/big-read/big-read-buy-local-what-price-costlier-singapore-produce-could-hamper-quest-greater-food-security-3552611.

[550] *Ibid.*

[551] *Ibid.*

[552] Tan, C. (2024a, April 7). Delays to opening of Singapore's 4th egg farm, which was slated to begin operations in 2024. The Straits Times. https://www.straitstimes.com/singapore/delays-to-opening-of-singapore-s-4th-egg-farm-that-was-slated-to-begin-operations-in-2024.

[553] *Ibid.*

[554] *Ibid.*

[555] Tan, C. (2024a, April 22). VertiVegies scraps plan to construct one of S'pore's largest indoor veggie farms. The Straits Times. https://www.straitstimes.com/singapore/vertivegies-scraps-plans-to-construct-one-of-s-pore-s-largest-indoor-veggie-farms.

[556] *Ibid.*

Feeding the Future with Innovation and Inclusivity

Singapore has approved 16 insect species for human consumption, including crickets, grasshoppers, mealworms, and silkworms as part of its broader strategy to enhance food security and diversify protein sources.[557] The United Nations has identified insects as a sustainable source of protein capable of supporting a global population projected to reach 9.7 billion by 2050.[558] In light of increasing food security challenges caused by climate change and geopolitical conflicts, interest in the nutritional and economic benefits of edible insects has grown.[559] Despite their high protein content and environmental sustainability, public acceptance remains limited due to cultural perceptions.[560] Efforts are underway to educate consumers and normalize insect consumption through awareness campaigns and culinary innovation.[561] Some restaurants in Singapore have begun incorporating insects into their menus; for instance, House of Seafood serves dishes like fish-head curry topped with crickets and tofu garnished with edible bugs, appealing to diners seeking unique gastronomic experiences.[562] "Vertical farming for these insects is good, especially organic. Whatever vegetables are left in my restaurant, I can feed them," said Francis Ng, founder of House of Seafood.[563] In accordance with food safety regulations, all insects approved for human consumption in Singapore must be farmed in controlled environments and cannot be harvested from the wild or fed with contaminants such as manure or spoiled food.[564]

In December 2020, Singapore became the first country to approve the sale of cultivated meat, granting regulatory approval to Eat Just's

---

[557] Kok, X. (2024, July 30). Crickets get crunchy as Singapore approves edible insects amid food security push. Reuters. Retrieved April 17, 2025, from https://www.reuters.com/world/asia-pacific/crickets-get-crunchy-singapore-approves-edible-insects-amid-food-security-push-2024-07-30/.

[558] *Ibid.*

[559] Wei, C. X. (2024b, July 8). Singapore Food Agency approves 16 insect species for consumption. The Business Times. https://www.businesstimes.com.sg/singapore/singapore-food-agency-approves-16-insect-species-consumption

[560] *Ibid.*

[561] *Ibid.*

[562] *Ibid.*

[563] Kok, X. (2024, July 30). *Op.cit.*

[564] *Ibid.*

cultivated chicken product sold under the label Good Meat.[565] This landmark decision by the Singapore Food Agency (SFA) positioned the country at the forefront of food innovation.[566] Since then, the SFA has approved additional cultivated meat products, including Vow's cultivated quail, and is expected to approve offerings from other international companies such as Vital Meat and Meatable.[567]

Singapore's regulatory framework emphasizes both safety and innovation, allowing companies to develop and market cultivated meat responsibly.[568] For example, GOOD Meat received approval to produce cultivated chicken using serum-free media, enhancing both scalability and ethical production standards.[569]

Complementing its progress in cultivated meat, Singapore is also fostering the growth of plant-based protein alternatives.[570] A notable example is Karana, a local startup that uses jackfruit to develop whole-plant-based meat substitutes.[571] Karana's products, such as jackfruit-based shreds and mince, are minimally processed and designed for diverse culinary applications.[572] The company sources its jackfruit from smallholder farms in Sri Lanka, reinforcing its commitment to sustainability and agricultural inclusivity.[573] In its first phase, KARANA launched in Singapore with six leading restaurants, including Candlenut, Butcher Boy, Open Farm

---

[565] Southey, F. (2023, February 16). Dissecting cultivated meat regulation part 2: What's working in the US and Singapore, and what's not? FoodNavigator.com. https://www. foodnavigator.com/Article/2023/02/16/dissecting-cultivated-meat-regulation-part-2-what-s-working-in-the-us-and-singapore-and-what-s-not.

[566] *Ibid.*

[567] Ong, S. and Chandran, R. (2024, June 19). Singapore doubles down on lab-grown meat as Silicon Valley backs off. Rest of World. Retrieved April 18, 2025, from https:// restofworld.org/2024/lab-grown-meat-singapore/.

[568] Southey, F. (2023, February 16). *Op.cit.*

[569] *Ibid.*

[570] SGN. (2021, August 17). Meet Daniel Riegler, the Karana co-founder creating Asia's first plant-based meat alternative using jackfruit. Singapore Global Network. Retrieved April 18, 2025, from https://singaporeglobalnetwork.gov.sg/stories/business/meet-daniel-riegler-the-karana-co-founder-creating-asias-first-plant-based-meat-alternative-using-jackfruit.

[571] *Ibid.*

[572] *Ibid.*

[573] Karana. (2021, May 12). Introducing Karana | Eat Karana. Eat Karana. Retrieved April 18, 2025, from https://eatkarana.com/introducing-karana.

Community, Morsels, Atout, and Grain Traders, highlighting the city's concentration of talented chefs and its status as a global leader in food tech innovation, according to Co-Founder Blair Crichton.[574]

The Future of Agriculture in Singapore

Singapore, a small island nation with a land area of about 720 sq. km, faces unique challenges in ensuring its long-term food security due to its heavy reliance on food imports and limited arable land.[575] Singapore's heavy reliance on imported food makes it vulnerable to disruptions like climate change, supply chain issues, and geopolitical tensions.[576] The COVID-19 pandemic underscored the fragility of global food networks and accelerated efforts to localize food production in urban centers like Singapore.[577]

In response to these vulnerabilities, the Singapore government initiated a strategy aiming to locally produce 30% of the nation's nutritional needs by 2030.[578] This strategic vision not only addresses food security but also positions agriculture as a growth sector underpinned by technological innovation and sustainability principles.[579] The Fourth Industrial Revolution (4IR) has already influenced Singapore's agricultural transformation by introducing advanced technologies such as

---

[574] *Ibid.*

[575] Ministry of Foreign Affairs – Singapore. (n.d.). Towards a Sustainable and a Resilient Singapore. United Nations | Department of Economic & Social Affairs – Sustainable Development. Retrieved April 18, 2025, from https://sustainabledevelopment.un.org/content/documents/19439Singapores_Voluntary_National_Review_Report_v2.pdf.

[576] Food. (2024, December 23). Ministry of Sustainability and the Environment. Retrieved April 18, 2025, from https://www.mse.gov.sg/policies/food.

[577] Tan, A. (2020a, April 2). S'pore taking steps to ensure adequate food supply. The Straits Times. Retrieved April 18, 2025, from https://www.straitstimes.com/singapore/environment/spore-taking-steps-to-ensure-adequate-food-supply.

[578] Md Ali, N. H. (2023, June 12). The Big Read: Buy local but at what price? Costlier Singapore produce could hamper quest for greater food security. CNA. Retrieved April 18, 2025, from https://www.channelnewsasia.com/today/big-read/big-read-buy-local-what-price-costlier-singapore-produce-could-hamper-quest-greater-food-security-3552611.

[579] *Ibid.*

artificial intelligence (AI), the Internet of Things (IoT), and robotics into urban farming systems.[580]

However, as agriculture enters the era of the Fifth Industrial Revolution (5IR), a new paradigm is an emerging one that emphasizes human-centric, ethical, and ecologically regenerative practices.[581]

The future of agriculture in Singapore is increasingly characterized by aquaculture, various agricultural techniques, alternative proteins, and circular food systems that are designed to operate within the constraints of a dense urban landscape.[582] The agri- and aquaculture sectors widely agree that high-tech systems can efficiently produce food, with Mr. Ray Poh from Artisan Green highlighting the need for further research to lower input costs and enhance scalability, similar to Singapore's successful approach to water management.[583] The integration of digital twins of Controlled Environment Agriculture (CEA), virtual replicas of real-world systems utilizing real-time data and simulations, enhances performance and optimizes decision-making.[584] Additionally, innovations in alternative protein production, ranging from cultivated meat to insect farming, are opening new frontiers in sustainable food innovation.[585]

---

[580]LKH Precicon. Industry 4.0 is changing Singapore. Retrieved April 18, 2025, from https://www.precicon.com.sg/news/industry-news/industry-4-0-is-changing-singapore.

[581]Adegoke, K. A., Adesina, O. O., and Okon-Akan, O. A. (2022). *Sawdust-biomass based materials for sequestration of organic and inorganic pollutants and potential for engineering applications. In Current research in green and sustainable chemistry.* IntechOpen. https://doi.org/10.1016/j.crgsc.2022.100274.

[582]Singapore Food Agency. (2022, November 11). Food for Thought | A sustainable food system for Singapore and beyond. Default. Retrieved April 18, 2025, from https://www.sfa.gov.sg/food-for-thought/article/detail/a-sustainable-food-system-for-singapore-and-beyond.

[583]Loh, R. (2024g, October 21). With several farms closing or struggling to break even, what is the future for agriculture in Singapore? CNA. Retrieved April 18, 2025, from https://www.channelnewsasia.com/today/big-read/high-tech-low-returns-farming-4684566.

[584]David, I. (n.d.). How digital twins will enable the next generation of precision agriculture. The Conversation. https://theconversation.com/how-digital-twins-will-enable-the-next-generation-of-precision-agriculture-213711.

[585]Lurie-Luke, E. (2024b). Alternative protein sources: Science powered startups to fuel food innovation. *Nature Communications*, 15(1). https://doi.org/10.1038/s41467-024-47091-0.

Singapore Current Agricultural Landscape

Agriculture in Singapore has historically played a minimal role in the national economy due to limited land availability, rapid urbanization, and a focus on industrial and service-oriented sectors.[586] Only a minimal portion of Singapore's land is allocated for agricultural purposes, as most conventional farming activities have been replaced by industrial and residential developments over the past five decades.[587] Nevertheless, small-scale farming persists in regions housing a range of local vegetable, egg, and fish farms that contribute modestly to the domestic food supply.[588] Singapore's local agri-food sector, dominated by hen shell egg, vegetable, and seafood farms, contributed 31.9%, 3.2%, and 7.3% respectively to the nation's total food consumption in 2023, a proportion that has remained stable over the past three years.[589]

To optimize productivity within spatial constraints, Singapore's agricultural sector has increasingly shifted toward high-tech, urban-based farming solutions such as hydroponics, aquaponics, vertical farming, and climate-controlled greenhouses.[590] The Agri-Food Cluster Transformation (ACT) Fund is a $60 million initiative by Singapore to support local farms in adopting advanced, resource-efficient technologies and sustainable practices to boost productivity and strengthen food security.[591]

---

[586] Food and The City: Overcoming Challenges for Food Security. (n.d.). Centre for Liveable Cities. Retrieved April 18, 2025, from https://isomer-user-content.by.gov.sg/50/da33e51c-ad22-4d45-969d-319bc63dd0fc/food-and-the-city-overcoming-challenges-for-food-security.pdf.

[587] *Ibid.*

[588] Urban Redevelopment Authority. (n.d.). Love local – Your guide of exploring our own backyard. Retrieved April 18, 2025, from https://www.ura.gov.sg/-/media/Corporate/Resources/Publications/Books/Love-local-guide.pdf.

[589] Singapore Food Agency. (n.d.). SINGAPORE FOOD STATISTICS 2023. Retrieved April 18, 2025, from https://www.sfa.gov.sg/docs/default-source/publication/sg-food-statistics/singapore-food-statistics-2023.pdf.

[590] SG101. (2023, April 5). *Challenge accepted: Urban farming in Singapore.* https://www.sg101.gov.sg/resources/archives/challengeaccepted-urban-farming-in-singapore/.

[591] Agri-food Cluster Transformation (ACT) Fund. (2024, October 18). Default. https://www.sfa.gov.sg/recognition-programmes-grants/grants/agri-food-cluster-transformation-act-fund.

These financial mechanisms incentivize farms to adopt automation, LED lighting, smart sensors, and data-driven management systems that improve crop yields and reduce labor dependence.[592]

The redevelopment of the Lim Chu Kang region into a 390-hectare high-tech agri-food cluster is a flagship example of Singapore's commitment to transforming its traditional agricultural base into a technologically advanced food production hub.[593] This project aims to triple current output by incorporating shared facilities, robotics, and digital agriculture platforms to improve land use efficiency and environmental performance.[594] In parallel, urban rooftop farms such as City Sprouts and Citiponics operate in repurposed spaces atop commercial buildings, demonstrating the potential for agriculture to integrate seamlessly into the urban fabric.[595]

Singapore's agri-food ecosystem also includes research institutions, agritech accelerators, and academic partnerships that support the development of next-generation farming solutions.[596] Institutes such as the Singapore Institute of Food and Biotechnology Innovation (SIFBI) and the Centre for Urban Agriculture are conducting R&D in plant science, microbial technologies, and food systems resilience.[597] Collaborations

---

[592] Mok, W. K., Tan, Y. X., and Chen, W. N. (2020). Technology innovations for food security in Singapore: A case study of future food systems for an increasingly natural resource-scarce world. *Trends in Food Science & Technology*, 102, 155–168. https://doi.org/10.1016/j.tifs.2020.06.013.

[593] Tan, A. and Tan, C. (2022, October 2). Lim Chu Kang set to be redeveloped into high-tech agri-food cluster: SFA. The Straits Times. Retrieved April 18, 2025, from https://www.straitstimes.com/singapore/environment/lim-chu-kang-set-to-be-redeveloped-into-high-tech-agri-food-cluster-sfa.

[594] Singapore's emerging AgriTech ecosystem. (2020, January 16). UNDP. Retrieved April 18, 2025, from https://www.undp.org/policy-centre/singapore/blog/singapores-emerging-agritech-ecosystem.

[595] Thai, L. K. (2021, February 10). Food for Thought | Singapore: Food security despite the odds. Default. Retrieved April 18, 2025, from https://www.sfa.gov.sg/food-for-thought/article/detail/singapore-food-security-despite-the-odds.

[596] A*STAR. (2021, November 21). A*STAR Launches New Research Institute to Support Singapore's Food Innovation Ecosystem. Agency for Science, Technology and Research. Retrieved April 18, 2025, from https://www.a-star.edu.sg/News/astarNews/news/publicity-highlights/a-star-launches-new-research-institute-to-support-singapore's-food-innovation-ecosystem.

[597] *Ibid.*

between government agencies and startups are vital, with companies like Archisen leading the way in controlled environment agriculture by using Internet of Things (IoT) technology to produce pesticide-free vegetables year-round.[598]

Despite these advancements, the agricultural sector continues to face challenges such as high energy costs, limited economies of scale, and difficulties in attracting and retaining skilled labor.[599] Consumer preference for imported produce and lack of public awareness about the benefits of local food also hinder demand, reinforcing the need for stronger engagement and education initiatives.[600] Nevertheless, the groundwork laid through strategic planning and technology adoption positions Singapore's agricultural sector to play an increasingly vital role in national food resilience and sustainability goals.[601]

Technological Innovation and Urban Farming

Technology lies at the core of Singapore's strategy to overcome its geographical constraints and achieve sustainable food production within a dense urban environment.[602] Urban farming in Singapore has evolved into a high-tech endeavor, with precision agriculture, Internet of Things (IoT) devices, automation, and robotics enabling unprecedented levels of efficiency and control.[603] AI systems, supported by sensors, enable continuous monitoring and automated regulation of key parameters such as

---

[598] Begum, S. (2020, October 30). Urban farm leverages smart technology to produce highly nutritious and flavourful greens. The Straits Times. https://www.straitstimes.com/singapore/urban-farm-leverages-smart-technology-to-produce-highly-nutritious-and-flavourful-greens.

[599] Tan, C. (2024, May 20). S'pore vegetable, seafood production fell in 2023 due to construction challenges, inflation: SFA. The Straits Times. https://www.straitstimes.com/singapore/s-pore-vegetable-seafood-production-fell-in-2023-due-to-construction-challenges-inflation-sfa.

[600] Koh, D. (2024, November 21). Strategies for a more sustainable path toward Singapore's food production goals. RSIS – Nanyang Technological University. https://rsis.edu.sg/rsis-publication/rsis/strategies-for-a-more-sustainable-path-toward-singapores-food-production-goals/.

[601] *Ibid.*

[602] *Ibid.*

[603] Kok, M. A. (n.d.). Singapore's emerging AgriTech ecosystem. UNDP – Singapore Global Centre. https://www.undp.org/policy-centre/singapore/blog/singapores-emerging-agritech-ecosystem.

temperature, humidity, light spectrum, CO_2 levels, and nutrient delivery, to optimize plant growth, ensure quality control, and maximize yield per unit area without human intervention.[604]

Vertical farming has emerged as one of the most promising solutions for high-density, pesticide-free agriculture, utilizing stacked growing systems within buildings to reduce land dependency.[605] MEOD is an innovative urban farm in Singapore that produces safe, fresh, and sustainable vegetables using pesticide-free methods, relying on organic and biological pest control with beneficial insects like ladybugs and praying mantises to ensure high-quality yields.[606] Kai Hian explained that agriculture in Singapore primarily involves overcoming spatial limitations, while also addressing challenges such as restricted land availability, labor constraints, and climatic conditions like heat and humidity that can place stress on crops.[607] To address these issues, MEOD employs advanced technologies, including climate control systems and vertical aeroponics that efficiently deliver water and nutrients to crops.[608]

The integration of IoT technologies into farming systems enables real-time monitoring and control of environmental conditions through wireless sensors and cloud-based platforms.[609] These smart systems allow farmers to remotely adjust settings via mobile applications and receive alerts about abnormalities such as temperature fluctuations or pest outbreaks.[610] Additionally, AI algorithms are increasingly being used to

---

[604] Cengiz Kaya. (2025). Intelligent environmental control in plant factories: Integrating sensors, automation, and AI for optimal crop production. *Food and Energy Security, 14*(1), e70026. https://doi.org/10.1002/fes3.70026.

[605] Teng, P. and Kim, S. (2021). Vertical farms: Are they sustainable? In S. Rajaratnam School of International Studies, RSIS Commentary. https://www.rsis.edu.sg/wp-content/uploads/2021/06/CO21092.pdf.

[606] Farm. (2024, July 4). Singapore Food Agency. Retrieved April 18, 2025, from https://www.sfa.gov.sg/fromSGtoSG/farms/farm/Detail/meod.

[607] *Ibid.*

[608] *Ibid.*

[609] Prem Rajak, Abhratanu Ganguly, Satadal Adhikary, Suchandra Bhattacharya, Internet of Things and smart sensors in agriculture: Scopes and challenges. *Journal of Agriculture and Food Research*, https://doi.org/10.1016/j.jafr.2023.100776.

[610] Vijendra Kumar, Kul Vaibhav Sharma, Naresh Kedam, Anant Patel, Tanmay Ram Kate, Upaka Rathnayake, A comprehensive review on smart and sustainable agriculture using IoT technologies. *Smart Agricultural Technology*, https://doi.org/10.1016/j.atech.2024.100487.

predict crop cycles, detect diseases, and optimize resource allocation based on historical and real-time data.[611]

Automation technologies are also transforming labor-intensive tasks in agriculture, with farms deploying robotics for seeding, transplanting, harvesting, and packaging.[612] These systems help mitigate labor shortages, improve consistency, and reduce human error, contributing to long-term sustainability and scalability.[613]

Archisen designs and operates advanced systems to grow local produce in urban environments, using its expertise in crop management, engineering, and IoT to link multiple farms to a centralized intelligence platform for improved efficiency and profitability.[614] With support from A*STAR's technological expertise, Archisen accelerated the development of its smart farm management system, Croptron, as explained by co-founder Sven Yeo.[615]

In addition to commercial farms, community and promote farming initiatives supported by technology are helping to build food literacy and public support for local agriculture.[616] Singapore's schools are embracing innovation by transforming learning spaces and integrating artificial intelligence (AI) to enhance education.[617] Jing Shan Primary School, for instance, has introduced the "Sprout Spot," an indoor hydroponics system

---

[611] Annu Singla, Ashima Nehra, Kamaldeep Joshi, Ajit Kumar, Narendra Tuteja, Rajeev K. Varshney, Sarvajeet Singh Gill, Ritu Gill, Exploration of machine learning approaches for automated crop disease detection. *Current Plant Biology*, https://doi.org/10.1016/j.cpb.2024.100382.

[612] Gordon-Smith, H. (2024, April 12). Exploring the Future of Agriculture: A Deep Dive into Robots. *Agritecture*. Retrieved April 18, 2025, from https://www.agritecture.com/blog/exploring-the-future-of-agriculture-a-deep-dive-into-robots.

[613] *Ibid.*

[614] A*Star. (2020, September 1). *Food Innovation for the Future – Collaborating for Sustainable living*. Agency for Science, Technology and Research. Retrieved April 18, 2025, from https://www.a-star.edu.sg/News/astarNews/news/features/food-innovation-for-the-future---collaborating-for-sustainable-living.

[615] *Ibid.*

[616] Woo Hoi Yuet. (2023, April 26). *Food security from the ground up: A community approach*. Gov Insider. Retrieved April 18, 2025, from https://govinsider.asia/intl-en/article/food-security-from-the-ground-up-a-community-approach.

[617] Tushara, E. (2024, November 14). Schools transform their spaces and learning experience with grant support and AI. The Straits Times. https://www.straitstimes.com/singapore/schools-transform-their-spaces-and-learning-experience-with-grant-support-and-ai.

where students cultivate vegetables like baby Bok choy and kale, promoting sustainability through hands-on experience.[618] This initiative is part of the school's Applied Learning Programme, emphasizing a "teach, embed, and live" approach to instill sustainable practices in students' daily lives.[619]

Singapore's robust digital infrastructure and proactive policy environment provide a conducive setting for technological experimentation in agriculture.[620] Agri-tech accelerators and innovation hubs, such as the Agri-Food Innovation Park in Sungei Kadut, bring together startups, research institutions, and corporates to test and scale emerging technologies.[621] As a global hub for business, innovation, and supply chains, Singapore hosts multinational companies, startups, and research institutions that drive the development and commercialization of cutting-edge solutions to address critical challenges in the Agri-Food sector.[622]

Despite significant investments and government support, many agri-tech ventures struggle with high operational costs.[623] The challenges of scaling technologically advanced farming methods highlight the need to reassess strategies to ensure the long-term viability and sustainability of high-tech agriculture in the country.[624] Therefore, ongoing investments in local R&D, workforce training, and resilient supply chains are essential to ensure the long-term viability and autonomy of high-tech urban agriculture in Singapore.[625]

Food waste valorization plays a key role in advancing the circular economy by converting discarded materials into valuable resources that

---

[618] *Ibid.*

[619] *Ibid.*

[620] Ministry of Foreign Affairs. *Op.cit.*

[621] Seow Bei Yi. (2019, March 5). Parliament: Agri-food innovation park in Sungei Kadut to open from early 2021, says Koh Poh Koon. *The Straits Times*. Retrieved April 18, 2025, from https://www.straitstimes.com/politics/parliament-agri-food-innovation-park-in-sungei-kadut-to-open-from-early-2021-koh-poh-koon.

[622] SIAW. (202 C.E., October 22). Singapore International Agri-Food Week 2024 to catalyse the transition to sustainable and nutritious food systems in Asia. Singapore International Agri-Food Week. Retrieved April 18, 2025, from https://sginternationalagrifoodweek.com.sg/singapore-international-agri-food-week-2024-to-catalyse-the-transition-to-sustainable-and-nutritious-food-systems-in-asia-2/.

[623] *Ibid.*

[624] *Ibid.*

[625] Food and The City: Overcoming Challenges for Food Security. *Op.cit.*

can be reintroduced into the supply chain to produce new goods.[626] Among various food waste streams, spent grains are the most successfully valorized, with approximately 82.7% (around 89,200 tons annually) being repurposed, likely for use as animal feed or in bioenergy production.[627] Meat waste also shows a high valorization rate of 70.1%, possibly through rendering processes or conversion into pet food.[628] Similarly, coffee waste achieves a valorization rate of 75.9%, with potential applications in composting, biochar production, or as a substrate for mushroom cultivation.[629]

Several start-up companies, including Insectta and Inseact, are utilizing black soldier flies (BSF) to recover nutrients from food waste and by-products of the food industry.[630] Regulatory constraints currently limit these companies to using primarily homogeneous food waste streams. However, in 2021, Blue Aqua received approval to process mixed food waste sourced by an airline company.[631] The company plans to convert this waste into insect protein for aquaculture by cultivating crickets and mealworms.[632]

Moreover, Singapore has taken a leading role in advancing future foods by investing approximately 24 times more public funding into protein innovation relative to its gross domestic product than the United States and other tech hubs, while also establishing a comprehensive safety approval framework for novel foods.[633] Innovate 360 is Singapore's first food accelerator that combines venture capital funding with comprehensive infrastructure to support agri-food and food-tech startups.[634] In

---

[626] Food waste valorisation. (n.d.). https://www.nea.gov.sg/our-services/waste-management/3r-programmes-and-resources/food-waste-management/food-waste-valorisation.

[627] *Ibid.*

[628] *Ibid.*

[629] *Ibid.*

[630] Mubita, T., Appelman, W., Soethoudt, H., and Kok, M. (2021). Resource and water recovery solutions for Singapore's water, waste, energy, and food nexus. Part II, Food waste valorization. https://doi.org/10.18174/554531.

[631] *Ibid.*

[632] *Ibid.*

[633] Gosker, E. M. and Zhou Weibiao. (2024, September 9). Science Talk: Supercharging "future foods" in Asia and beyond. The Straits Times. https://www.straitstimes.com/singapore/science-talk-supercharging-future-foods-in-asia-and-beyond.

[634] Innovate 360. (2024, August 15). Innovate 360 – Accelerator with Facilities & Venture Capital. Innovate 360 – Singapore's Food Accelerator With Facilities. https://

addition to funding and infrastructure, Innovate 360 provides mentorship from over 30 industry experts to help startups scale across Asia and globally, supporting ventures in deep tech, agri-tech, sustainability, and consumer goods, and is recognized by APEC as a key food technology partner in advancing food security.[635] These centers serve as platforms for collaboration between researchers, corporations, and policymakers, driving forward the translation of R&D into scalable commercial applications.[636]

Vision Beyond 2030

The emergence of the Fifth Industrial Revolution (5IR), which emphasizes human-centric design, ethics, and ecological regeneration, is expected to shape the next phase of agricultural development in Singapore.[637] Unlike the Fourth Industrial Revolution (4IR), which emphasized automation and digitalization through technologies such as artificial intelligence, the Internet of Things (IoT), robotics, additive manufacturing, and data analytics to transform manufacturing and supply chains, its core focus lies in leveraging smart, connected systems to improve efficiency, productivity, flexibility, and sustainability across production processes.[638]

The Fifth Industrial Revolution (5IR) integrates social values and sustainability into technological progress, making it especially relevant for food systems by enabling the creation of more personalized products, enhancing worker well-being, and fostering sustainable practices that generate value not only economically, but also socially and environmentally.[639]

Singapore is well-positioned to be a leader in this movement due to its existing infrastructure in agri-tech, food science, and sustainability

---

innovate360.sg/.

[635] *Ibid.*

[636] Innovate 360. (2024, August 15). *Op.cit.*

[637] What is Industry 5.0? (Top 5 Things You Need To Know). (n.d.). https://www.twi-global.com/technical-knowledge/faqs/industry-5-0.

[638] Kaempf, D. (2025, April 16). Singapore's Journey Towards Industry 4.0. Team-Metal Precision Machining Contract Manufacturer. https://teammetal.com/singapores-journey-towards-industry-4-0.

[639] What is Industry 5.0? (Top 5 Things You Need To Know). *Op.cit.*

governance.[640] Ongoing investment in bio-innovation will empower the nation to develop scalable, nutritious, and environmentally regenerative food solutions.[641] Research institutions like the Singapore Institute of Food and Biotechnology Innovation (SIFBI) are already leading efforts to create new food materials with low resource footprints.[642] In parallel, Singapore's higher learning institutions play a crucial role in advancing agri-food innovation through education, research, and partnerships.[643] The National University of Singapore (NUS),[644] Nanyang Technological University (NTU),[645] and Singapore Institute of Technology (SIT) contribute significantly through specialized programs in food science, biotechnology, and environmental sustainability.[646] Institutions such as the Singapore University of Technology and Design (SUTD),[647] Republic Polytechnic,[648] Temasek Polytechnic,[649] and Ngee Ann Polytechnic are also key players, offering training and applied research in smart farming,

---

[640] A*STAR. (n.d.). Creating Growth, Enhancing Lives. Annual Report, April 2022 – March 2023 https://www.a-star.edu.sg/docs/librariesprovider1/default-document-library/annualreports/annualreport2023.pdf.

[641] *Ibid.*

[642] A*STAR. (2021, November 21). *Op.cit.*

[643] EDB Singapore. (2023, December 5). Singapore Climate Change Initiatives & Ecosystem for Sustainability. Retrieved April 19, 2025, from https://www.edb.gov.sg/en/business-insights/insights/singapores-climate-action-ecosystem-for-fostering-sustainable-development.html.

[644] National University of Singapore. (n.d.). Department of Food Science and Technology. In National University of Singapore [Report]. https://www.science.nus.edu.sg/wp-content/uploads/2022/02/NUS-CHS-FST-Brochure.pdf.

[645] Food Science & Technology. (n.d.). School of Chemistry, Chemical Engineering and Biotechnology (CCEB). https://www.ntu.edu.sg/cceb/fst.

[646] SIT. (2024, March 22). Supporting the Nation's Sustainable Development Plans. Singapore Institute of Technology. Retrieved April 19, 2025, from https://www.singaporetech.edu.sg/news/supporting-nations-sustainable-development-plans.

[647] Making artificial intelligence work for sustainability – Singapore University of Technology and Design (SUTD). (2024, October 8). Singapore University of Technology and Design (SUTD). https://www.sutd.edu.sg/news-listing/making-artificial-intelligence-work-for-sustainability.

[648] Agriculture Research and Innovation (AGRI) Centre. (n.d.). AGRI. https://www.rp.edu.sg/agri.

[649] Agri-Food Technology | Temasek Polytechnic. (n.d.). Temasek Polytechnic. https://www.tp.edu.sg/landing/industry-partners/agri-food-technology.html.

aquaculture, and sustainable food production.[650] Additionally, the Institute of Technical Education (ITE) supports workforce development through hands-on training in vertical farming, hydroponics, and agri-tech operations.[651] Together, these academic and technical institutions form a comprehensive ecosystem that reinforces Singapore's capacity to become a regional leader in future food systems.[652]

Although emerging food sources and production systems are still in their early stages, they hold significant potential to address global food security challenges more sustainably in the long term.[653] For example, alternative proteins such as cultivated meat and fermentation-derived proteins can be produced using comparatively less land and labor.[654] However, these innovations present unique challenges, particularly in ensuring safety and establishing effective regulatory frameworks.[655] Deputy Prime Minister Gan Kim Yong announced that the Singapore Food Agency (SFA) is developing proposals at the Codex Alimentarius Commission to standardize safety assessments for cultivated meat, aiming to streamline regulatory approvals and build consumer trust, and called on global food regulators to support this initiative during his address at the Singapore International Agri-Food Week on 18 November 2024.[656]

Complementing these efforts, Singapore-based research labs are also advancing sustainability in conventional staples, such as rice; one such initiative that develops climate-resilient rice varieties and integrates smart farming practices was recently recognized by the World Economic Forum for its potential to reduce emissions and enhance water efficiency in rice

---

[650] Industry services. (n.d.). Ngee Ann Polytechnic (NP). https://www.np.edu.sg/schools-courses/academic-schools/school-of-life-sciences-chemical-technology/industry-services.

[651] Fully automated facility at ITE to train students in high-productivity urban farming. (2022, November 11). The Straits Times. https://www.straitstimes.com/singapore/ite-s-new-semi-automated-facility-to-train-students-in-high-productivity-urban-farming.

[652] Pursuing careers in the Agri-Food sector. (2024, October 16). Default. https://www.sfa.gov.sg/farming/careers-in-agri-food-sector/pursuing-careers-in-agri-food-sector.

[653] Food for Thought | Singapore: Food security despite the odds. (2024, June 5). *Op.cit.*

[654] DPM Gan Kim Yong. (2024, November 18). DPM Gan Kim Yong at the Singapore International Agri-Food Week 2024. Prime Minister's Office Singapore. Retrieved April 19, 2025, from https://www.pmo.gov.sg/Newsroom/DPM-Gan-Kim-Yong-at-the-Singapore-International-Agri-Food-Week-2024.

[655] *Ibid.*

[656] *Ibid.*

cultivation.[657] Temasek Life Sciences Laboratory's (TLL) "Decarbonising Rice" initiative, which aims to reduce methane emissions, conserve water, and increase rice yields, won the Giving to Amplify Earth Action (GAEA) award in the Climate, Nature and Resilience Science category at the World Economic Forum in January.[658]

A similar initiative to the Singapore lab's sustainable rice cultivation is the collaborative effort between Singapore and the Philippines to develop healthier and more climate-resilient rice varieties.[659] This partnership focuses on creating rice strains that are resistant to droughts and floods and have a lower glycemic index, addressing both environmental challenges and health concerns, such as the rising prevalence of diabetes in Asia.[660] The collaboration involves Singapore's Temasek Life Sciences Laboratory and the International Rice Research Institute (IRRI), aiming to enhance food security and nutritional outcomes in the region.[661]

Reflection Notes

Singapore's agricultural transformation is a compelling narrative of resilience, innovation, and strategic foresight. Despite severe spatial limitations and an overreliance on imported food, the city-state has redefined its agricultural identity through policy integration, technological advancement, and a systems-thinking approach. The evolution from colonial-era gambier and nutmeg plantations to cutting-edge vertical farms and climate-controlled greenhouses highlights how environmental and geopolitical constraints have acted not as obstacles but as catalysts for innovation. This transformation is not merely about increasing food production; it represents a fundamental reimagining of agriculture as a sophisticated,

---

[657] Chin Hui Shan. (2025, March 24). Project by Singapore lab to cultivate rice sustainably wins WEF award. The Straits Times. Retrieved April 19, 2025, from https://www.straitstimes.com/singapore/initiative-by-singapore-lab-to-cultivate-rice-sustainably-wins-wef-award.

[658] *Ibid.*

[659] Raguraman, A. (2024, August 17). Singapore looking to develop healthier rice varieties with the Philippines. The Straits Times. https://www.straitstimes.com/singapore/singapore-looking-to-develop-healthier-rice-options-with-the-philippines.

[660] *Ibid.*

[661] *Ibid.*

high-tech, and knowledge-driven sector embedded within an urban ecosystem.

One of the most striking features of Singapore's agri-food strategy is how innovation is woven into its policy architecture. Initiatives such as the "30 by 30" goal, the Agri-Food Cluster Transformation (ACT) Fund, and the Singapore Food Story R&D Programme reflect a governance model that does not merely react to global food security challenges but proactively builds structural resilience. This institutional support is further complemented by a vibrant entrepreneurial ecosystem, where start-ups like GroGrace, Sustenir, and Comcrop operate at the intersection of technology, sustainability, and localized food production. In Singapore, innovation is not an isolated effort confined to laboratories, but a culture deeply embedded in public-private collaboration and urban planning.

Yet, amid these advances, it is vital to acknowledge that sustainability is not solely a technological pursuit. While data-driven indoor farms and controlled-environment agriculture (CEA) promise resource efficiency and year-round productivity, the ethical dimensions of food systems remain equally important. Singapore's agricultural journey provokes deeper questions about equity, sovereignty, and ecological stewardship. The Anthropocene, the current epoch in which human activity significantly alters Earth's systems, compels societies to develop food systems that are not only productive but also regenerative and inclusive. In this light, Singapore's approach offers not just a model for efficiency but a platform to explore ethical food futures grounded in responsibility and interconnectedness.

Looking outward, Singapore's agri-tech innovation offers scalable lessons for other urbanized and resource-scarce regions. These lessons include the importance of systems integration through linking policy, education, infrastructure, and enterprise, and the value of designing circular systems that minimize waste and optimize resource use. Moreover, Singapore demonstrates how localization of food production, when paired with digital innovation, can significantly enhance national food resilience. As global food systems grow increasingly vulnerable to climate change, pandemics, and geopolitical instability, Singapore stands out as a living testbed for what future-ready agriculture might look like.

As the world moves toward what may be characterized as Agriculture 5.0, an era defined by regenerative practices, ethical technologies, and ecological restoration, Singapore must continue to evolve. While Agriculture 4.0 has centered on automation, connectivity, and data optimization, the next frontier demands deeper integration with nature, circular economy principles, and even biotechnological innovations such as fungi-based materials and soil-less cultivation systems. The farms of the future will not merely be efficient; they will need to be symbiotic, drawing on ecological intelligence as much as artificial intelligence.

Ultimately, Singapore's agricultural journey is about more than food production; it is about transformation. It reflects how a nation can turn its limitations into opportunities by aligning innovation with sustainability and policy with purpose. In a world increasingly shaped by ecological disruption and supply chain fragility, Singapore's model offers not only practical insights but also philosophical guidance. Food systems should nourish both people and the planet. This case study is thus both a reflection and a roadmap. It illuminates the challenges of today while pointing toward a regenerative and equitable future for urban agriculture in the decades to come.

The next chapter of the volume focuses on Godo's case studies of Japan.

Chapter 2

Japan's New Model of Farming Under the Industrial Revolution 4.0

Yoshihisa Godo

Abstract

Godo's chapter begins with a brief overview of industrial revolutions and how they have transformed Japan's industries and agriculture. This useful backgrounder then proceeds to explain how this industrial process transformed the agricultural industry from one of comprehensive life-style materiel provision to more focused food materials. Kenji Maeda's herb farming is featured as one example of the balance between these two objectives and goals, something made possible with the use of net-based awareness/knowledge dissemination solutions to ICT (Information and Communications Technology) tutorials.

Introduction to scientific farming in Japan. This sub-sectional case study discusses a new type of "smart agriculture". It studies how farmers like Maeda created a new business model based on the growing new communication technology. This chapter will discuss a new type of "scientific farming".

This was followed by a chapter that first provided the historical background behind Japan's agricultural land reforms and detailed how land reforms may benefit from Industry 4.0. Godo argued that, unlike mechanization before Industry 4.0, artificial intelligence (AI), combined with information technology (IT), enables flexible thinking that can make more flexible and dynamic use of agricultural land possible.

As with the previous chapters, Godo provides background information on the geographical specs of Japan's marine spaces. He then proceeds to detail the modern features of Japan's fisheries and fishing industries. The coverage is detailed, even touching on work ethics, culture, and legislative acts (including examples of deregulations). This chapter advocates how updated legislations have made it possible for IoT and AI, as well as other Industry 4.0 technologies can make it possible for efficient record-keeping, big data generation, and maintenance of databases possible for all functions, from monitoring water condition to protection of endangered species.

Introduction

The role of agriculture has changed over time. Before the First Industrial Revolution (approximately 1760–1830 in England, then worldwide), agricultural products were used not only as food but also for various necessities like clothes, cosmetics, decorations, furniture, musical instruments, stationery, and toys (e.g., animal skin, beeswax, Chinese indigo, cotton, feather, jute, mat rush, silk, and wool).

Today, agricultural products are primarily seen as food materials, indicating that the manufacturing sector has largely replaced agriculture in producing non-food items.

So far, the manufacturing sector has not surpassed agriculture in food production. However, with advancements in biotechnologies and information technologies under Industry Revolution 4.0, AI-equipped robots in fully automated factories could potentially outperform farmers in food production. This raises the question: Should farmers disappear from the world? This study examines the evolving role of agriculture in the context of Industry Revolution 4.0.

The Division of Labor and the First Industrial Revolution

Before discussing Industrial Revolution 4.0, it is useful to review the First Industrial Revolution.

The First Industrial Revolution marked the beginning of continuous high-speed economic growth (in per-capita GDP), originating in the UK from the late 18th century to the early 19th century, then spreading to continental Europe by the mid-19th century.

Key innovations during this period included the mule spinning machine and the steam engine. However, all inventions eventually wear out. Despite this, per-capita GDP has continued to grow since the First Industrial Revolution, highlighting the mechanism behind this sustained growth.

Adam Smith's *The Wealth of Nations*, published in 1776, effectively describes this mechanism. Smith argued that a laborer could produce only 20 pins per day if managing the entire production process alone. However, if the process is divided into 18 distinct operations, with each assigned to specific workers, over 4,000 pins can be manufactured per laborer. This significant increase in productivity is known as the benefit of the division of labor.

The division of labor creates new businesses. For example, if the process of making screws is spun out as a new business field, the screw industry will start training expert laborers specialized in producing screws. In economics textbooks, these are called "externality", which promotes further increase in GDP by activating the overall economy. A further increase in GDP will promote the further development of the division of labor. As such, the division of labor will create a virtuous circle in the economy.

"The Wealth of Nations" contrasts sharply with Malthus's "An Essay on the Principle of Population." Malthus assumes that the economy follows the Law of Diminishing Returns (LDR), meaning total production (or GDP) cannot grow faster than the total population.

Ricardo, a notable opponent of Malthus, also supported the LDR but emphasized the role of physical capital (e.g., machine tools) in production. He distinguished between two sectors: the modern sector (non-agriculture) and the traditional sector (agriculture). Ricardo argued that the modern sector would grow through physical capital accumulation, whereas the traditional sector would be constrained by the LDR due to its reliance on natural resources, such as cultivable land. Consequently, Ricardo expected unavoidable food shortages and a halt to overall economic growth.

Contrary to the pessimistic views of Malthus and Ricardo, per-capita GDP has increased since the First Industrial Revolution, indicating that the LDR is incorrect. Despite this, mainstream economists have continued to advocate for the LDR, likely to attract public attention by emphasizing a sense of impending crisis.

Alfred Marshall's "Principles of Economics," published in 1890, shifted mainstream thinking. Instead of the LDR, Marshall introduced the Law of Increasing Returns (LIR), proposing that per-capita GDP should continue to rise. The vicious cycle of division of labor supports the LIR. Additionally, Marshall refined Adam Smith's argument using the economics framework of "Industrial Organization Theory." His work marked a turning point in academia, resulting in the LIR being recognized as the standard vision for economic growth.

The First Industrial Revolution and the Commodification of Labor Power

Around the mid-19th century, a new production system, known as the "American System of Manufactures" (ASM), emerged, characterized by the use of interchangeable components. In producing polyethylene terephthalate bottles, for example, caps and bodies are made separately, allowing any cap to fit any body. Companies prepare machine tools and instruction manuals for each operation.

Under the ASM, laborers must meet two conditions: (1) possess sufficient scientific knowledge to understand instructions, and (2) be accustomed to group behavior at workplaces with the inorganic rhythm of the clock. These conditions render laborers interchangeable, leading to the phenomenon known as the "commodification of labor power" (CLP).

The ASM originated in the US during the mid-19th century, primarily in the manufacturing of weapons and automobiles, triggering the global development of the CLP across various industries.

The CLP differs from humans' natural (or pre-modern) aptitude and behavior. Hence, modern society requires systems to instill these conditions in children, leading to the emergence of the modern education system. Notably, modern education is physiologically abnormal, as school-aged children, who have high physical activity levels and low disease resistance, are in cramped and unhealthy classrooms.

This indicates that the primary purpose of today's education system is to promote economic growth. If the Industrial Revolution 4.0 alters humanity's role in the economy, the education system must also undergo reform.

Replacement of Labor by Machines After the First Industrial Revolution

In economics textbooks, commodities or services whose demand increases faster than income are termed "luxuries." Leisure is one such luxury. Consequently, labor supply per capita decreases as per-capita income rises (note that economic models assume a person's time is divided between labor, which is necessary for income, and leisure). This dynamic increases the wage rate in the economy.

In response to wage increases following the First Industrial Revolution, labor-saving technologies like mechanization were developed. Machines excel at repetitive tasks, easily replacing simple labor jobs. In contrast, intellectual roles, such as inventions and technological advancements, cannot be replaced by machines.

This situation heightens the importance of tertiary education. In primary and secondary education, children learn group activities and basic-level science, which is sufficient for blue-collar jobs, which are more vulnerable to automation. Therefore, to maintain labor opportunities, workers should pursue further academic training through tertiary education.

Agriculture and the Division of Labor

Per-capita GDP has shown an upward trend since the First Industrial Revolution, as seen in Figure 1. This implies that Marshall's view of the LIR is correct, while Malthus's view of the LDR is incorrect. Today, mainstream economists argue that the vicious cycle of the division of labor, along with the CLP, is the main source of modern economic growth. This argument appears persuasive for modern industries, such as factories and supermarkets, developed after the First Industrial Revolution.

However, the effectiveness of the division of labor in agriculture is questionable. Agriculture differs from modern industries in three key ways: (1) The "main actors" in agriculture are crops and livestock, with farmers as "supporting actors," while human beings are the main actors in modern industries. (2) The primary energy source in agriculture is solar rays, accessible at no cost, whereas modern industries rely on man-made power, such as electricity, which comes with costs. (3) Agricultural

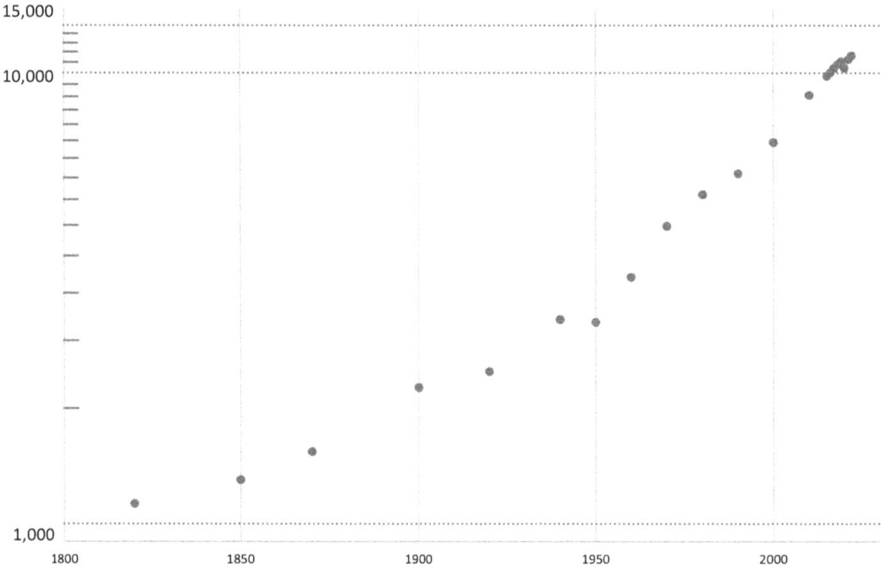

Figure 1. Real GDP per capita of the world economy in 2011\$ (semi-logarithmic graph).

Source: Maddison Project Database 2023, https://www.rug.nl/ggdc/historicaldevelopment/maddison/releases/maddison-project-database-2023.

production conditions change unpredictably, while modern industrial conditions can be controlled by humans (for example, solar rays fluctuate with the weather, while electricity intake can be adjusted according to operational schedules).

In farm management, reducing damage from abnormal growth, such as disease, is a top concern. It is impossible to completely block the risk of abnormal growth due to the nature of animals and plants. However, by providing appropriate treatments as soon as a farmer detects signs of abnormality, damage can be minimized. Careful and consistent observation of crops and livestock is necessary. Farming involves various tasks: plowing, pumping and draining water, raising seedlings, weeding, spraying fertilizers, and harvesting. By performing all these tasks themselves, farmers continue to improve their skills. Therefore, the division of labor is unprofitable in farming. Despite this, the division of labor persists in agriculture, as farmers are instilled with the aptitude for the CLP during their childhood in school.

In developed countries, where the division of labor in agriculture is highly advanced, farmers struggle to maintain their farms without government protection. Due to their strong political influence, governments continue to supply subsidies to farmers.

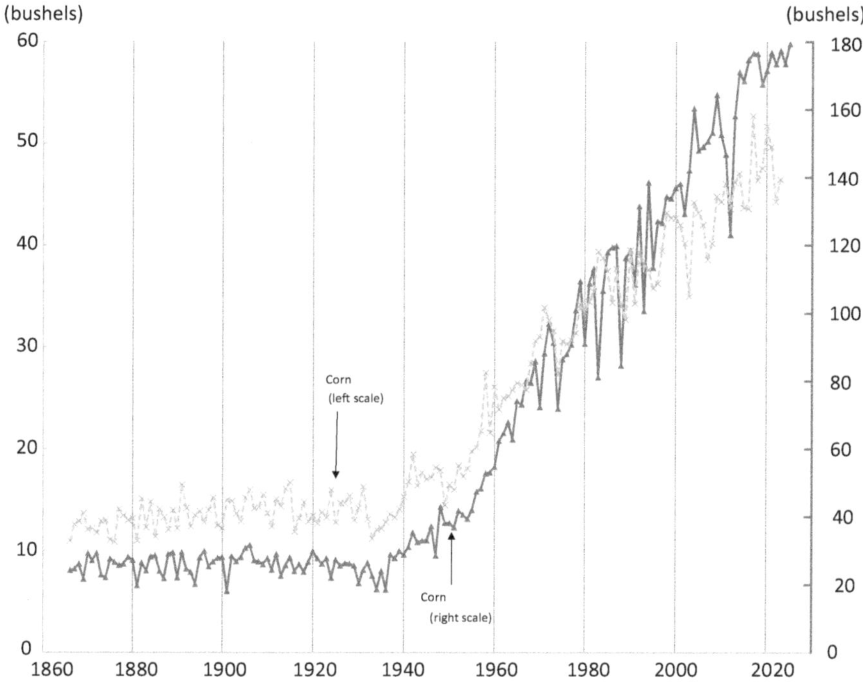

Figure 2.   Yield of acre of corns and wheat in the US.

Source: Martin and Brokken (189, p. 159); Luttrell and Gilbert (1976, p. 527), USDA National Agricultural Statistics Service, Historical Data On Line, 2024.

As illustrated in Figure 2, yield per acreage sharply increased nearly one hundred years ago, driven by the use of underground resources in farming. During the First World War, developed countries invested in scientific research to create new military technologies. This led to the development of new agricultural inputs, such as chemical fertilizers and tractors, derived from military innovations like poisons and tanks. These inputs rely on underground resources such as iron and oil, with oil also powering agricultural machinery. Over the past century, ongoing advancements in agricultural technologies have heavily relied on these resources.

In summary, agricultural production in developed countries has increased due to government subsidies and the exploitation of underground resources, such as oil. Consequently, developed countries have become major crop exporters, resulting in a declining trend in international crop prices over the past century, as indicated in Table 1 and Figure 3.

Table 1. Global supply-demand conditions for grains (million tons).

	1961–1963 average			1979–1981 average			1999–2001 average			2010-2012 average		
	Output	Consumption	Net export	Output	Consumption	Net export	Output	Consumption	Net export	Output	Consumption	Net export
World	855	855	0	1,511	1,511	0	2,060	2,060	0	3,025	3,025	0
Developed countries	283	287	–3	516	476	39	637	530	107	681	608	74
Japan	20	23	–3	14	36	–12	12	39	–26	11	36	–25
USA	152	138	14	294	226	68	335	255	80	382	322	60
Others	111	125	–14	208	214	–17	290	237	53	288	250	39
Developing countries	572	569	3	996	1,034	–39	1,424	1,530	–107	2,344	2,418	–74
Middle-income countries	263	258	5	418	449	–31	484	565	–81	1,133	1,194	–61
Low-income countries	309	310	–2	577	585	–8	939	965	–26	1,211	1,223	–12

Notes: (a) Grains include barley, maize, millet, oats, rice, rye, sorghum, and wheat, (b) "Consumption" is calculated by subtracting net export from production, (c) Developed countries are OECD member economies in which the 1998 GNI per capita was $9,360 or more. Middle-income countries are economies in which the 1998 GNI per capita was between $761 and $9,361, and non-OECD member economies in which the 1998 GNI per capita was $9,360 or more. Low-income countries are economies in which the 1998 GNI per capita was $760 or less. These income criteria are same as those set by the World Bank (*World Development Indicators*, 2000), (d) "World" is obtained by aggregating "Developed countries" and "Developing countries." Countries whose data are unavailable are excluded from the calculation (thus, "World" in this table does not exactly match the FAO's estimates for the world total), (e) "Net export" in low-income countries is calculated from those in developed and middle-income countries (so that "net export" in the world total equals zero), (f) There are few summation errors because of rounding, (g) Including Japan among the developed countries, the figures for the developed countries are obtained.

Source: FAO, *FAOSTAT Database* (2000, 2004, 2016).

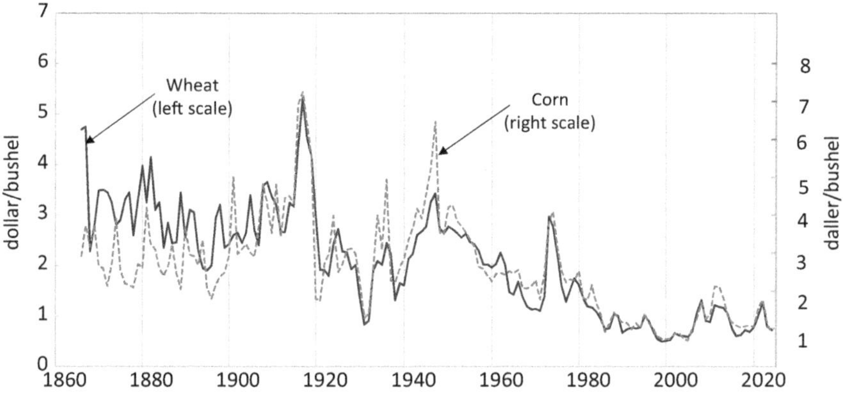

Figure 3.   Long-term changes in real prices (deflated by 1967 CPI).

Source: Martin and Brokken (1983). USDA National Agricultural *Statistics Service, Historical Data On Line (2008); World Bank*, Commodity Markets (online) 2023; International Monetary Fund, *World Economic Outlook Databases*, Hayami and Godo (2005).

Table 2.   Energy consumption per capita (tons).

	Maddison estimates	Malanima estimates
1820	0.21	0.22
1900	0.52	0.55
1950	0.84	0.89
2003	1.71	1.67

Note: Animal power is not included.
Source: Maddison (2007) and Malanima (2020).

It is important to note that the intensive use of underground resources extends beyond agriculture. Fossil energy plays a crucial role in nearly all economic activities. Table 2 illustrates the rapid growth of per-capita energy consumption since the First Industrial Revolution, with fossil energy accounting for approximately 80% of total energy consumption during the period covered in Table 2.

Recently, wind and solar power have been promoted as alternative energy sources that may reduce reliance on fossil energy. However, these technologies also rely on underground resources, such as rare metals. Thus, even if alternative energy sources replace fossil fuels, society will remain over-reliant on underground resources. This over-reliance poses a significant problem for future generations, although the exhaustion of these resources is unlikely to occur within the present generation; therefore, this study does not further address the issue of exhaustion.

Potentials of Cell-Based Food

Housing, food, and clothing are the basic necessities of life. Today, agriculture is viewed as the industry that produces food materials. Historically, agricultural products served as raw materials not only for food but also for housing and clothing. Items like beeswax, cotton, hemp, jute, rush, silk, livestock skin, and wool were popular materials for clothing and housing.

The First World War was epoch-making. European countries rushed to develop military equipment, facilities, and weapons, such as uniforms, parachutes, and poison gases. Initially intended for military use, these innovations led to the development of new consumer products, including nylon, alloyed steel, and plastic, which were so useful and inexpensive that they replaced traditional agricultural products in clothing and housing. Consequently, the agricultural sector focused primarily on food production.

A new type of manufactured commodity, cell-based food, may soon replace agricultural products in food production. Cell-based food is developed by culturing animal and/or plant cells in controlled factory conditions, free from the effects of climate change and disease.

Currently, the technologies for cell-based food are experimental. One obstacle to its commercialization is the discomfort it evokes in consumers. Many governments restrict cell-based food technologies to laboratories and prohibit practical use, likely due to their eerie perception. Most people, including researchers and politicians, are hesitant to promote cell-based food.

In the near future, AI is expected to play a crucial role in national economic decisions, including budget allocation and food safety policy-making. AI, free from subjective human feelings, may accelerate the promotion of the cell-based food industry to practical application.

AI and Fake Utopia

The development of AI technologies will drastically change human life-styles. AI excels at intellectual jobs and is projected to surpass human intelligence by 2045[1]. This suggests that robots equipped with AI will outperform humans in intellectual jobs and farming, areas previously considered difficult to automate until the advent of Industrial Revolution 4.0. Even those with tertiary education may struggle to find job opportunities. In this scenario, people would not need to work for income; instead, everyone would receive a basic income and focus solely on consumption, leading to a life free from hard work—a so-called "fake utopia."

In the fake utopia, how should people live? What is the social role of agriculture? In seeking answers to these questions, Kenji Maeda's herb farming offers important insights.

Herb Farming in Japan

The following part of this chapter discusses Kenji Maeda's unique herb farming. Before detailing Kenji Maeda's case, a brief review of the history of herbs in Japan is useful. Globally, an herb is defined as "a plant whose leaves, flowers, or seeds are used for flavor, medicine, or pleasant smell"[2]. However, in Japan, the word "herb" is usually used in a narrower sense, referring only to exotic ingredients, such as basil, lavender, rosemary, and thyme. In this study, the word herb refers to this notion.

For years, herbs were unfamiliar to most Japanese. This changed in 1985 when Seiko Hirota popularized herbs in Japan. As the main instructor of a popular TV program *"Shumito Engei"* (literally translated as "Hobby and Gardening"), she demonstrated how to enjoy various types of herbs and wrote numerous articles and books on the subject. As a consequence, herbs became widely recognized as new plants with an elegant appearance and a pleasing aroma. Now, herb saplings are popularly sold at home centers in Japan, which Japanese people enjoy growing them at home.

Kenji Maeda, who resides in Matsue, Japan, introduces a novel method of farming, advocating that customers should enjoy growing herbs at home. He supplies healthy herb saplings and offers tutoring services.

Kenji Maeda has a 600-square-meter farm for herb saplings and runs a virtual shop, "Soramimi Herb Shop," online. He receives orders for

---

[1] Ray Kurzweil, *The Singularity Is Near: When Humans Transcend Biology*, Viking Adult, 2005.

[2] Oxford University Press, *Oxford Advanced Learner's Dictionary*, 10th edition, 2020.

saplings and provides tutoring through the shop's website, with trade and communication mainly occurring via the Internet.

Currently, annual sales of saplings exceed 20 million yen (approximately US$140,000). He employs four part-time workers (including a developmentally disabled person) and two full-time workers (including himself). Kenji Maeda's herb farming is a successful, profitable business that creates jobs for his neighbors. How did Kenji Maeda establish this farming style? What can we (businesspersons, politicians, and researchers) learn from his case? By answering these questions this study aims to introduce a novel concept for the future agricultural industry.

Kenji Maeda's Trials and Errors

Kenji Maeda was born in 1969 in Matsue, a small town in western Japan, Shimane Prefecture. After graduating from high school, he entered Kobe University to study psychology. He became interested in the psychological analysis of the healing effects of aroma and started growing herbs in his apartment. With no prior experience, he read several horticulture books, which sparked a deeper interest in herbs than in psychology. He decided to train as an herb expert, but needed more space to grow them. Farm rents in Kobe were too high, but his parents' community network helped him secure a leased farmland in Matsue. In 1991, he quit Kobe University, moved back to Matsue, and began growing herbs.

Kenji Maeda planned to open a shop to sell aroma products and herb saplings, but lacked initial capital. He accumulated money by taking various jobs, including maintenance services at a nature park.

During this time, he traveled to Yamanashi Prefecture for professional training with Makoto Abe, a leading herb expert in Japan, which significantly improved his farming skills and knowledge.

In 1995, Kenji Maeda opened Soramimi Herb Shop (meaning "air and ear") in Matsue, selling herbal products, accessories, ornaments, tea leaves, and herb saplings. Despite expectations, the shop's performance was poor, generating almost no profits. Kenji Maeda believed Matsue's population of nearly 200,000 was too small to attract customers. In 2007, he launched an online section for the shop, targeting consumers across Japan. However, website sales totaled less than 10,000 yen per year, insufficient for a sustainable living. Consequently, he took on various jobs, including public park maintenance, which further hindered his business. Over time, his experience growing herbs on leased farmland boosted his confidence in his knowledge and skills.

In 2012, Kenji Maeda attended a training course on website design organized by the Shimane Industrial Promotion Foundation (SIPF), which consisted of 10 lessons (each nearly five hours long) taught by Kanji Fukuhara, a renowned website designer. This was Kenji Maeda's first formal training in creating a customer-attractive website; previously, he built the Soramimi Herb Shop website using self-taught skills.

Through this training, Kenji Maeda realized that his business's poor online sales were due to two main issues: confusing product listings that obscured the shop's concept and numerous obstacles in the purchasing process.

In the same year, applying lessons from SIPF, Kenji Maeda redesigned the website and converted Soramimi Herb Shop into a purely virtual store, focusing solely on herb saplings.

This marked a turning point for Kenji Maeda's business, leading to an unprecedented surge in orders for herb saplings, reaching nearly 6,000 in 2020. Nearly half of his current customers reside in Tokyo and nearby prefectures, where high population density limits gardening space, allowing only for small plants.

Kenji Maeda's New Model of Farming

Since herbs are relatively new to Japan, most Japanese people lack sufficient knowledge about them. Many Japanese households have limited living space, preferring to grow decorative plants, including herbs, in planters on verandas instead of gardens. Consequently, their way of using herbs differs from that of North Americans and Europeans.

Consider lavender: it blooms early in summer, and people enjoy its fragrance in their homes during this season. Therefore, home centers sell the most lavender saplings in April, displayed in disposable plastic cups for easy transport. However, transplanting saplings in April is inappropriate as they may become weak and die within a month or two. Thus, although lavender is a perennial herb, it is often treated as a short-term decorative item, similar to a cut flower. Japanese horticulturists' practices for growing lavender differ from those in North America and Europe. To efficiently use limited farmland, many Japanese horticulturists aim to grow lavender quickly, employing various chemicals, fertilizers, and heat sources. As a result, the saplings often lack strong immunity, but this does not concern households that purchase them as short-term decorations.

Kenji Maeda's approach to enjoying lavender differs from the popular Japanese pattern. He sells healthy herb saplings with strong immunity, believing customers should nurture them long-term. He collects lavender cuttings in April and grows them into saplings for eight months without chemicals or pesticides. In November, considered the best month for transplanting, he offers these saplings on his website, priced more than twice that of those at ordinary home centers, reflecting his confidence in their quality.

The SIPF training he underwent in 2012 taught him the importance of high-quality website design. He regularly improves and updates the Soramimi Herb Shop website.

Kenji Maeda advises customers on caring for herbs at home and enjoys personally tutoring them. The Internet allows him to exchange online videos, photos, and messages, sharing the joy of nurturing herbs.

Herbs, easily grown in planters, are ideal for Japanese novices in plant cultivation. Consequently, most of his customers lack experience with herbs or any plants. On his website, Kenji Maeda explains meticulous herb care, utilizing photos and videos. He builds personal relationships with online customers by collecting digital photos of their herbs for a photo contest on the shop's website and providing updates about his farmland.

Any customer can receive Kenji Maeda's personal tutoring on how to grow herbs by using ICT (such as sending digital photos and/or videos of the current situation of herbs through the Internet). Each day, Kenji Maeda does such personal tutoring for nearly 20 customers. After personal tutoring, he revises the Q&A section of the shop's website to enhance it, which helps attract new customers.

Kenji Maeda only sells herb saplings, a rare farming style in Japan. His farming exemplifies a successful niche strategy. Although his herb saplings are priced higher than those at home centers, customers are willing to pay for the tutoring service he offers.

Table 3 presents the income and expenditures of Soramimi Herb Shop for 2016–2021. Despite the small farmland area of 600 square meters, the farm generates over 20 million yen in revenue. The COVID-19 pandemic created a surge in demand for herb saplings, resulting in a sharp increase in sales in 2020. Sales then declined slightly from 2020 to 2021, as pandemic restrictions eased, although Japan still had some movement limitations in 2021.

Table 3.   The income and expenditures of Kenji Maeda's herb farming (thousand yen).

Fiscal year	2012	2013	2014	2015	2016	2017	2018	2019	2020	2021
Total revenue (gross sales)	9,347	11,041	12,314	14,941	15,053	14,472	16,207	17,748	21,379	20,304
Total expenditures	6,705	5894	8,044	9,935	11,137	11,448	12,664	13,054	15,652	13,694
Personnel expenses	2,338	1,681	2,695	3,599	3,841	3,931	3,769	4,288	4,726	3,872
Packing and shipping expenses	1,740	1,947	2,404	2,886	2,893	2,782	3,838	4,177	5,176	4,764
Miscellaneous expenses	1,277	1,780	2,129	2,337	2,655	2,533	3,635	2,489	3,957	2,423
Communication expenses	765	365	572	764	730	709	699	771	887	992
Rent for land	120	120	120	120	120	120	120	120	120	120
Taxes and public charges	5	5	5	5	5	415	496	48	22	237
Net income	2642	5,147	4,271	5,006	3,915	3,023	3,544	4,693	5,727	6,610

Notes: (a) The Japanese fiscal year starts on April 1 and ends on March 31 next year, (b) Personel expenses include salaries, bonuses, and allowances.
Source: Keiji Maeda's documents for tax payment.

Labor costs include payments to employed laborers only, excluding imputed costs for Kenji Maeda's own labor. The table shows that packing and shipping costs are the largest portion of total expenditures due to the delicacy of herb saplings, which require special protective materials for safe delivery.

Kenji Maeda's herb farming model minimizes cash outflow by personally procuring the necessary agricultural supplies. For instance, he collects fowl droppings, procures leaf mold compost from neighbors at low or almost zero cost, and prepares fertilizers himself.

Concluding Remarks

Agriculture is typically defined as the production of commodities for daily consumption (e.g., food) or industrial use (e.g., cocoon, mat-rush). Kenji Maeda has created a model focused on the joy of growing plants, primarily targeting beginners who purchase herb saplings from his website and receive care instructions from him.

In developed countries like Japan, people are generally satisfied with their clothing and food, resulting in limited demand for additional commodities. Instead, those who are materially fulfilled often seek new, non-material joys, which Kenji Maeda helps to provide. He also developed a form of smart agriculture, where information and communications technology is used to enhance customer communication rather than traditional farming processes.

In the future, unmanned operations and advancements in AI and ICT may replace farmers in food production. In this context, nurturing plants at home could become a popular leisure activity, creating a new business opportunity for farmers. If this occurs, Kenji Maeda will be recognized as a "pioneer" in this type of farming.

The next chapter zooms in on how one of the major foundations of the agricultural industry, land use, can benefit from Industry 4.0 technologies.

Industry 4.0 and Japan's Agricultural Land-Use System

Failure in agricultural land-use planning is among the biggest problems currently facing Japanese agriculture. Owing to the lack of clear zoning regulations, agricultural and residential land uses are intermingled. Often, artificial lights from houses impede the growth of crops, and residents become allergic to chemicals sprayed on agricultural land. Agricultural landowners commonly leave their agricultural land idle without considering the negative effects on neighboring farmers (pests may proliferate in idle land).

This does not mean that Japan lacks experience in designing and implementing a system for effective agricultural land-use planning. By contrast, up until the Meiji period, Japan's farmers had well-organized agricultural land use. Furthermore, during the early post-Pacific War period, agricultural cooperatives successfully functioned as coordinators of agricultural land use.

Why does Japan currently lack effective agricultural land-use planning? What is the Japanese government's response to this problem? Will Industry 4.0 provide a solution? By answering these questions, this chapter discusses the changing structure of Japan's agricultural sector.

Democracy and Land-Use Planning

Since the Meiji Restoration in 1868, Japan has sought to emulate various social systems from European and North American countries. The land-use system was no exception. Thus, before discussing Japan's agricultural land-use problem, it is useful to briefly review the land-use planning systems in European and North American countries.

The respect for private property rights, developed during the democratization process in European and North American countries, is among the most important aspects of modern society. However, respect for private property rights does not imply that owners can use their properties unconditionally. Owners' rights are restricted if they conflict with public interest.

Land use is a typical case in which private property rights should be restricted to protect public interest. For example, if a slaughterhouse is built next to a hospital, patients may not recuperate effectively. Similarly, if all the houses in a community are built to preserve the landscape, all residents and visitors will benefit from it.

Designing and enforcing a land-use plan poses a challenge. Land conditions differ widely by area; therefore, uniform regulations by the central government are inefficient. European and North American countries have invented various systems for citizens' participation in land-use planning, such as initiatives and referenda in the United States and *enquete* public enquiries in France. The coordination between the private ownership system and the restriction of private rights to land use through land-use planning is key to maintaining comfortable lives for citizens.

Land-use problems are closely related to democracy. Democracy comprises two components: the assertion of private rights and participatory democracy, namely, the participation of citizens in local government. As part of participatory democracy, citizens are obliged to participate in land-use planning. Similarly, citizens are allowed to trade and use land freely, provided they comply with the land-use plan.

The United States has typical cases of participatory democracy in land-use planning: local citizens accept responsibility for participating in land-use planning. Intensive discussions occur in public meetings, such as the City Planning Council. Local citizens are also engaged in the implementation of city planning. Citizens monitor the implementation strictly to ensure compliance with the plan.

Japan adopted democracy under the supervision of the General Headquarters of the Allied Forces (GHQ) during the early post-Pacific War period. As discussed later in this chapter, Japanese citizens' behavior since then suggests that they have neglected their responsibility in participatory democracy while entirely embracing the assertion of private rights. This may be because asserting private rights was—and remains—easier to emulate than practicing participatory governance.

Japan's Geography and Agricultural Land-Use Planning

Japan's agricultural land-use problem is closely related to its geography. Japan experiences high rainfall levels. Annual precipitation in Japan averages 1728 mm, more than twice that in France (750 mm), which is one of the largest agricultural countries in the European Union.

Japan is a mountainous country. Only one-third of the total land acreage in Japan is flat, which is much smaller than that in other major developed countries (e.g., 70% in the United States, 90% in the United Kingdom, 77% in Italy, 73% in France, and 69% in Germany).

Rivers in Japan are steep and short, rainfall water flows quickly into the ocean, and river water levels vary widely and continuously depending on the weather. Moreover, all farmers in a community use the same irrigation channels. Farmers must collaborate in the irrigation and extermination of pests; this collaboration among farmers is particularly important in rice farming. When rice paddy fields are flooded for irrigation purposes, water moves from the upper to the lower parcels.

Inappropriate water use on agricultural land in one parcel can adversely affect all farmers in the community.

To prohibit inappropriate water use on agricultural land, land-use planning at the community level is necessary. Land-use details, such as the rotation of water intake and drainage and the scheduling of planting and harvesting, must be carefully coordinated among farmers. Agricultural land-use planning was among the most important tasks for rulers of Japanese society because, until the Pacific War, the agricultural sector comprised the largest share of economic activity in the country.

Agricultural Land-Use Planning Before the Pacific War

Despite Japan's long history of agricultural land-use planning, participatory democracy is largely neglected in current land-use planning.

During Japan's feudal period, the Tokugawa shogunate (1603–1868) established a remarkably stable agricultural land-use system, which enabled them to retain power for more than two and a half centuries. The shogunate owned all the agricultural land, and transactions of agricultural land among farmers (such as buying, selling, lending, or borrowing) were prohibited. Only the eldest son of a farm household could succeed to his father's right to farm. The Tokugawa shogunate allocated land-management rights to feudal lords called *Daimyo*. *Daimyo* were responsible for constructing common agricultural assets, such as irrigation facilities, selecting crops, and designing and enforcing agricultural land-use planning. In return, *Daimyo* were authorized to collect agricultural land tax from farmers.

As *Daimyo* lived in cities, they faced difficulty monitoring farmers' behaviors. Therefore, *Daimyo* appointed village headmen to manage land-use planning and land tax collection at the local level. Moreover, farmers were organized into five-household neighborhood units that shared collective responsibilities for farming according to *Daimyo*'s land-use plans, and they were required to pay agricultural land tax to *Daimyo*.

Researchers have concluded that the Tokugawa shogunate system had several disadvantages. These include inhumane restrictions on fundamental human rights, particularly freedom of occupation choice and mobility. However, it has been widely recognized that agricultural land was used in a well-planned manner under this system.

The end of the Tokugawa shogunate and the inauguration of the Meiji government (in 1868) had a significant impact on the agricultural land-use system. To counter the colonization pressure by Western countries, the Meiji government sought to strengthen Japan's economy and modernize Japanese society. To achieve this, the government introduced a free-market system. As part of this system, in 1873, the government launched the Land Tax Reform (LTR) and introduced a private land ownership system (both agricultural and non-agricultural land). In the LTR, the government established a taxable value for each land parcel and issued land certificates to parcel owners. Land certificate holders were obligated to pay an annual asset tax equivalent to 3% of the assessed taxable value. Land became tradable through the sale and purchase of land certificates.

After the LTR was instituted, wealthy local families became large landowners by purchasing agricultural land in their villages. Before the 1920s, the Japanese financial market was still underdeveloped and regionalized, and wealthy local families preferred to remain in their villages and preserve their wealth primarily through agricultural land ownership.

As the use of labor-saving agricultural technologies such as harvesters and pesticides was limited during this period, the optimal farm size was as small as one hectare. Thus, wealthy local families farmed only a limited proportion of their agricultural land, and most of the land was leased to numerous small tenant farmers in their villages. These families engaged in daily farming activities; therefore, as "self-farming landlords," they possessed abundant knowledge about local agriculture. They supported tenant farmers' farming to increase rental yields. They also enjoyed high social standing in their hometowns because of their political and economic leadership. Accordingly, they were actively involved in the development of communal agricultural infrastructure, such as joint irrigation facilities. As local agriculture leaders, they also engaged in designing and enforcing agricultural land-use planning in the communities.

Consequently, agricultural land-use planning was conducted relatively smoothly under the leadership of wealthy local families. However, their leadership was not organized under written law. As long as the families maintained strong ties with their tenant farmers, they preserved community order under their leadership. However, the relationship between the families and tenant farmers was vulnerable to shifts in economic and political conditions. In this sense, the leadership of local wealthy families in agricultural land-use planning was fragile and unstable in the long term.

As Japan's development stage shifted from light to heavy industrialization around 1920, the economic concerns of wealthy local families changed. They began investing in new factories in urban areas for greater financial returns and became involved in the management of heavy industry enterprises. Even after these families moved to urban areas, they retained agricultural land as part of their portfolios. In contrast to self-farming landlords, the so-called "absentee landlords" neither possessed deep knowledge about local agriculture nor assumed leadership in the development of communal agricultural infrastructure. As the heavy industry sector expanded in urban areas, the number of self-farming landlords decreased while that of absentee landlords increased.

Conflicting interests existed between absentee landlords and tenant farmers. For example, absentee landlords often demanded higher rent from tenant farmers, leading to strong resistance and fierce village-wide disputes that were referred to as "peasant disputes." Peasant disputes began in the second half of the 1910s in western Japan, where heavy industrialization had begun, and the disputes spread nationwide during the 1920s. At their peak, the number of peasant disputes reached 7,000 annually, indicating that farmers were unable to focus on farming. As such, the absence of agricultural land-use planning became a serious problem in Japan in the late 1920s.

Ironically, militarism provided a solution to this problem. Japan shifted to wartime footing in the 1930s, and private rights were severely restricted. Approval from prefectural governors became mandatory for any alteration in agricultural land use.

The government established a new nationwide system for farmers, called *Nogyokai*. All farmers in each municipality were mobilized to join *Nogyokai*'s municipality office.

The prefectural governors appointed local leaders as administrative board members of *Nogyokai*'s municipality offices, tasking them with maintaining community discipline to support the wartime economy. Agricultural land-use planning was designed and enforced under the leadership of *Nokyokai*'s municipality offices.

During the Pacific War, the government imposed a strict rationing system on both agricultural inputs (such as fertilizers) and products (such as rice). *Nogyokai*'s municipality offices were involved in this system, serving as the sole channels through which farmers could obtain agricultural inputs and ship agricultural products.

Under this rationing system, the government determined the details of trade conditions, such as prices, at every distribution stage. All agricultural landowners were obliged to ship all agricultural products to the government through *Nogyokai*. In procuring agricultural products, the government applied different prices based on whether the agricultural landowner was self-farming, offering higher prices for self-farming. This was because the government could not afford to pay much money to non-self-farming landowners. This price differentiation implied that the government favored self-farming even before Japan's defeat in the Pacific War.

Agricultural Policy Reform Under the GHQ

In 1945, Japan unconditionally surrendered to the Allied Forces and came under the governance of the General Headquarters of the Supreme Commander for the Allied Powers (GHQ). To democratize Japan's rural communities, the GHQ mandated the Japanese government to implement Agricultural Land Reform (ALR), under which agricultural land owned by absentee landowners—and any land exceeding one hectare owned by resident landowners—was transferred to tenant farmers.

In 1946, each municipality established a special administration committee called the Agricultural Land Committee (ALC) to implement the ALR. The ALC comprised representatives of tenant farmers and agricultural landowners, elected from among these groups. It determined which parcels of agricultural land would be transferred from agricultural landowners to tenant farmers, as well as the cost of these transfers. Generally, the ALC sets agricultural land prices at extremely low levels. In most cases, the cost of 900 kilograms of rice was enough to transfer the ownership of 1 hectare of agricultural land from an agricultural landowner to a tenant farmer, despite the average yield of rice per hectare being nearly 5,000 kilograms.

From 1947 to 1950, 1.9 million hectares of agricultural land owned by 2.4 million landowners were transferred to 4.8 million tenant farmers.[3] Consequently, small-scale (approximately 1.0–1.5 hectares) owner farming became dominant nationwide.

The Agricultural Land Act was established in 1952 as a new, basic legal framework for agricultural land-use regulations.

---

[3] According to the Statistical Survey of Crops, the total acreage of Japan's agricultural land in 1950 was 5.0 million hectares.

Two Legal Frameworks for Leasing Agricultural Land

In 1951, the ALC was transformed into an Agricultural Committee (AC). The AC comprised two types of members: elected and appointed. Members were elected from among farmers and appointed by the municipal mayor. The total number of AC members differed by municipality (typically ranging between 10 and 35). Elected members had to outnumber the appointed members.

The Agricultural Land Act (ALA) stipulated that changes in the ownership and use of agricultural land required the AC's approval. Therefore, the AC held an important position in the implementation of the ALA, as any new transaction involving ownership changes was critical.

To prevent former agricultural landowners from repurchasing agricultural land parcels that were already transferred to tenant farmers during the ALR, the ALA provided strong legal protection to protect the interests of tenant farmers. For example, the ALA forbade landowners from cancelling a leasehold contract without a tenant farmer's consent.

Due to the ALA's strong protection for tenant farmers, agricultural landowners became reluctant to enter into new leasehold contracts with tenant farmers. Consequently, illegal leasehold of agricultural land (i.e., leasehold on informal agreements between agricultural landowners and tenant farmers without reporting to the AC) increased. This pressured the Ministry of Agriculture, Forestry and Fisheries (MAFF) to create a new legal framework for agricultural land leaseholds in 1975: The Program for the Promotion of Agricultural Land Use (PPALU). This program allowed agricultural landowners and tenant farmers to establish leasehold contracts flexibly. Although leasehold contracts based on the ALA remained available, those based on the PPALU became much more popular among both agricultural landowners and tenant farmers. In 1992, the PPALU was renamed Program for the Promotion of Agricultural Management Framework Reinforcement (PPAMFR).

Establishment of Agricultural Cooperatives

When the GHQ began ruling Japan, the treatment of *Nogyokai* was a major problem. The GHQ regarded *Nogyokai* as a militaristic system and sought to dissolve all its offices. Simultaneously, it required new democratic institutions to carry out the food rationing system in place of *Nogyokai*. In 1947, the Agricultural Cooperatives Act was established, and accordingly, *Nogyokai's* municipality offices were transformed into agricultural cooperatives.

During their establishment, agricultural cooperatives inherited most of the physical and human resources of *Nogyokai*'s municipality offices. However, they also differed from *Nogyokai* in two aspects. First, the president of an agricultural cooperative is elected by a vote among regular members. Second, farmers have no legal obligation to join agricultural cooperatives.

Agricultural cooperatives provide not only agricultural services but also a wide range of everyday services, including banking, supermarkets, and arrangements for family ceremonies. Each agricultural cooperative has its own non-overlapping jurisdiction, and the vast majority of farmers are regular members of agricultural cooperatives in their respective jurisdictions. Non-farmers who pay dues are allowed to use all services as associate members, and regular and associate members have equal rights to the use of these services. However, only regular membership holders possess voting rights at general meetings, in which agricultural cooperatives make the most important decisions about their activities.

When agricultural cooperatives began, Japan was experiencing a severe shortage of food and fertilizers. Thus, as successors to *Nogyokai*, these cooperatives also rationed commodities. Consequently, all farmers joined agricultural cooperatives, despite not being legally obligated to do so.

As discussed previously, the ALR led to the rapid emergence and dominance of similar small-scale owner farmers in rural areas. In this new context, agricultural cooperatives provided a suitable framework for farmers with shared interests to collaborate. These cooperatives organized various joint farming activities, such as maintaining communal agricultural warehouses and cleaning irrigation channels.

Although the rationing system was abolished in the early 1950s, agricultural cooperatives maintained collaboration with the MAFF to implement the MAFF's policy.[4] In return, the MAFF granted them preferential treatment, including a dominant role in fertilizer distribution to farmers and rice collection from farmers.

The Ministry of Finance (MOF) also treated agricultural cooperatives favorably. Until the 1990s, the MOF's control over financial institutions (including agricultural cooperatives) was so strong that the Japanese

---

[4]The Ministry of Agriculture, Forestry and Fisheries was named the Ministry of Agriculture and Forestry until 1978. For simplicity, this paper refers to it as the Ministry of Agriculture, Forestry and Fisheries for the period before the renaming.

financial sector was often referred to as "armed convoys." The MOF's permission was required for activities such as determining branch locations and creating new financial instruments. Simultaneously, interest rates were maintained below market levels, and the entry of new financial institutions was restricted, preserving the profitability of existing market participants. Agricultural cooperatives were granted greater freedom than ordinary commercial banks to establish branches, enabling them to generate profits by lending through the interbank money market to city banks that faced substantial financial demands from large businesses. Additionally, the agricultural cooperatives' term-deposit interest rate was 0.1 percentage points higher than the rate set by commercial banks, giving them a distinct advantage in attracting businesses.

Since then, the Liberal Democratic Party (LDP), which was founded in 1955 and has remained almost entirely in power, has established strong connections with agricultural cooperatives. Several characteristics of agricultural cooperatives make them particularly attractive to LDP politicians. First, farmers live in the same villages for generations and are accustomed to collective actions under the guidance of agricultural cooperatives. Second, in the multiple-member constituency system, which was maintained for Japan's general election (election of the members of the House of Representatives) until 1994, agricultural cooperatives are good at coordinating and allocating farmers' votes among plural LDP politicians in a constituency (note that, in order to keep a majority in the House of Representatives, LDP need to win plural sheets in a constituency). As agricultural cooperatives supported the LDP in elections, the LDP considered the agricultural cooperatives' interests when drawing up bills for the House of Representatives.

Based on their strong ties with the MAFF and the LDP, agricultural cooperatives often functioned as de facto sub-government bodies. A typical example was the acreage control program: the rice production cartel organized by the MAFF. This program was launched in 1971, when old rice stock accumulated to an unsustainable level, and continued until 2004 to regulate rice production. The MAFF first set a national target for the acreage that should be diverted from rice planting, allocating this target among cities, towns, and villages. In both formulation and allocation stages, the MAFF considered the opinions of agricultural cooperatives. Each agricultural cooperative prepared an agricultural land-use plan in its own jurisdiction according to the allocated target acreage. As the timing of water drainage and intake differs between rice farming and other crops, agricultural cooperatives carefully zoned agricultural land in rural communities into rice-farming and other areas. Furthermore, they monitored whether farmers followed their agricultural cooperatives' agricultural land-use plans for the acreage control program. As such, many agricultural policies would not have been executed without the help of agricultural cooperatives; therefore, the MAFF and the LDP did not introduce policies without considering agricultural cooperatives' reactions.

Protection and Regulations Regarding Agricultural Land Use

After the Pacific War, motorization increased non-agricultural land use (for houses, public facilities, and shopping centers). This created a new problem for Japan's land use, namely, which agricultural land should be converted for non-agricultural purposes and which should be protected from conversion.

"High-quality agricultural land' is defined differently in Japan than in other developed regions (Australasia, North America, and the European Union). A key distinction is that in Japan, 'high quality for farming' implies 'high potential for non-agricultural purposes.' The favorable conditions for modern farming are a flat terrain, abundant sunlight, conveniently sized and regularly shaped plots, adequate water supply and drainage, and good access to roads. Ironically, these five conditions also favor the conversion of agricultural land for non-agricultural uses, such as the construction of shopping centers and public facilities.

As agricultural land has various externalities, such as flood prevention, numerous statutes prescribe protections and regulations governing its use. Among these, the Law Concerning the Construction of Agricultural Promotion Areas (LCAPA) is particularly significant. The LCAPA authorizes municipal governments to designate the zoning of exclusively agricultural areas (EAAs). Farmers in EAAs are required to use agricultural land only for farming purposes. Abandonment of farming and conversion to non-agricultural uses are prohibited in EAAs. In return, agricultural land in EAAs receives favorable treatment in taxation and allocation of agricultural subsidies. Furthermore, the MAFF undertakes extensive agricultural land improvement investments in EAAs. These investments increase not only agricultural productivity but also the potential for agricultural land conversion.

On the surface, strict written laws govern agricultural land use. However, the implementation of these laws has been problematic. Agricultural land-use regulations are often manipulated if authorities face strong political pressure. An agricultural landowner's "desirable" scenario is that, under normal circumstances, their agricultural land is included in an EAA, enabling them to enjoy agricultural subsidies, the MAFF's agricultural land improvement investments, and reduced property taxes. Thereafter, when an agricultural land conversion plan (such as for a shopping center or a public construction project) is established, the agricultural land is excluded from the EAA, and local authorities promptly approve the conversion, allowing the landowner to realize substantial capital gains. Such "desirable" scenarios frequently materialize. Consequently, land use becomes increasingly haphazard and inefficient, disrupting both the agricultural and non-agricultural sectors. As mentioned previously, agricultural cooperatives were efficient in designing and implementing agricultural land-use plans. However, they were (and remain) uninvolved in the decisions regarding agricultural land conversion. This was (and remains) a fundamental limitation of agricultural cooperatives as organizational actors.

Declining Organizational Ability of Agricultural Cooperatives

The Japanese government launched drastic reforms to its financial market and electoral system. Although the government's aim in implementing these reforms was not related to agriculture, they had a substantial negative impact on agricultural cooperatives. As agricultural cooperatives were more strongly protected than other financial institutions under the MOF's armed convoy-style policy, financial liberalization in the 1990s significantly reduced the profitability of their financial activities. In fact, agricultural cooperatives were on the verge of bankruptcy in 1996, after their non-performing loans to so-called *jusen* (housing financing companies) reached intolerable levels. Although they survived the crisis through a 685 billion yen bailout led by the MOF, their financial condition has remained weak ever since.

In the 1994 electoral reform of the House of Representatives, the multiseat system was replaced with a single-seat system. Simultaneously, by increasing the number of seats in urban areas and reducing those in rural areas, the disparity in the value of votes between urban and rural areas decreased. Consequently, agricultural cooperatives lost their power as voting groups.

Owing to these reforms in the 1990s, agricultural cooperatives were unable to maintain their leadership in organizing farmers. The ties between them and the MAFF also weakened. Accordingly, the acreage-control program was abolished in 2004.

Manufacturers' Requests for Agricultural Trade Liberalization in the 1980s And 1990s

Manufacturers' groups such as *Keidanren* (the Japanese abbreviation of the Japan Federation of Economic Organizations) and *Doyukai* (the Japanese abbreviation of the Japan Association of Corporate Executives) strongly support the LDP and significantly influence its policymaking.

Keidanren and *Doyukai* confronted agricultural cooperatives regarding agricultural policy during the General Agreement on Tariffs and Trade (GATT)'s Uruguay Round of trade negotiations (1986–1994). In international trade negotiations, Japan was requested to abolish its rice autarky policy, that is, the Japanese government's prohibition of rice imports despite domestic prices being far higher than international prices.

At that time, Japan's manufacturers, who benefited from strong international competitiveness, generated a large trade surplus. This invoked severe criticism of the Japanese government by the North American and European governments (and citizens), who were suffering from widespread factory closures and high unemployment. Japan's import regulations on agricultural commodities (including its rice autarky policy) were regarded as unfair trade practices in the GATT's Uruguay Round of trade negotiations. Fearing a further increase in anti-Japan sentiments worldwide, *Keidanren* and *Doyukai* argued that Japan should liberalize its rice imports.

In this argument, *Keidanren* and *Doyukai* advocated for the "theory of international division of labor" (TIDL), which is popular in economics textbooks. The TIDL claims that a country's GDP is maximized if the country concentrates on producing internationally competitive commodities under a free international market system. This means that Japan had to export manufactured commodities and import agricultural commodities.

As a counterargument against *Keidanren* and *Doyukai*, agricultural cooperatives asserted that agricultural commodities were daily necessities with a much higher social value than market prices; therefore, TIDL should not be applied to them. Furthermore, agricultural cooperatives argued that Japan should not increase its agricultural imports to protect itself from internal policy disturbances. This logic, called "food security theory," was emotionally appealing (and still appeals to some in the contemporary period) to the Japanese public.

During the GATT's Uruguay Round negotiations, major member countries' criticism of Japan's rice autarky policy mounted to a critical level, forcing Japan to open its rice market by setting a rice import quota (equivalent to 5% of Japan's total rice consumption).

Through debates on the rice autarky policy, *Keidanren* and *Doyukai* realized that Japanese consumers are strong supporters of domestic agriculture. This experience provides the background to *Keidanren* and *Doyukai*'s changing opinions on Japanese agriculture in the 2000s, as discussed in the next section.

New Concept of "Aggressive Agriculture"

In retrospect, the GATT's Uruguay Round of trade negotiations was the last period in which Japan's manufacturing sector was strong. In the middle of the 1990s, the Japanese economy experienced a large-scale credit crunch. This discouraged Japanese manufacturers from investing in new businesses and technologies. Consequently, they gradually reduced their competitive power in the international market. Japan's per capita gross domestic product (GDP) has stagnated for nearly 30 years since the middle of the 1990s. This period is commonly referred to as "Japan's lost 30 years".[5]

By contrast, GDP has increased rapidly in neighboring countries. For example, China achieved remarkable economic growth during Japan's lost 30 years, joining the WTO in 2001. This accelerated China's industrialization, making it the "world's factory."

Reflecting on the weakening international competitive power of Japanese manufacturers, following the GATT's Uruguay Round trade negotiations, *Keidanren* and *Doyukai* changed their opinions. They stopped referring to the TIDL. Instead, they became concerned about the withdrawal of government protection from Japanese manufacturers. However, *Keidanren* and *Doyukai* did not wish to declare publicly that the Japanese manufacturing sector had fallen behind in the international

---

[5] See Hoshi, Takeo, and Anil Kashyap, *Corporate Financing and Governance in Japan: The Road to the Future* (MIT Press, 2004).

market for manufactured commodities; such a declaration would have attracted negative public perceptions.

In 2004, the Japan Economic Research Institute, a private research institute founded and sponsored by major manufacturer groups including *Keidanren* and *Doyukai,* submitted a new idea for revitalizing Japanese agriculture, called "aggressive agriculture." The concept of aggressive agriculture can be summarized as follows:

(1) Japanese agriculture has significant potential.
(2) Current farming methods are overly outdated for realizing this potential.
(3) If non-agricultural companies begin farming, they introduce modern technology and management strategies to the agricultural sector.
(4) By combining agriculture and other businesses (e.g., green tourism, food processing, restaurants), the Japanese agricultural sector's profitability can increase. This combination, referred to as "the sixth industrialization," should be the key revitalization strategy.[6]
(5) Non-agricultural companies should advocate for and promote the sixth industrialization.
(6) The Japanese government should provide financial and institutional support for the entry of non-agricultural companies into agriculture and their involvement in the sixth industrialization process.

Keidanren and *Doyukai* recognize that the Japanese public tends to be tolerant of government policies regarding subsidy allocation and border protection for the agricultural sector. By promoting the concept of aggressive agriculture, *Keidanren* and *Doyukai* have used agriculture, including agriculture-related businesses, as a tool to campaign for the withdrawal of government support, including subsidies.

---

[6]Economics textbooks commonly classify production activities into three categories: primary industry (concerned with obtaining or providing natural raw materials for conversion into commodities and products for the consumer), secondary industry (converts the raw materials provided by the primary industry into commodities and products for the consumer), and tertiary industry (part of a country's economy concerned with the provision of services). The Sixth Industrialization implies that by combining agricultural products (commodities of the primary industry) with manufacturing, such as food processing (the secondary industry), and services, such as tourism (the tertiary industry), agriculture would be more profitable. This mechanism is compared to the mathematical equation primary (1) + secondary (2) + tertiary (3) = sixth (6). This is the name' origin, sixth industrialization.

As discussed earlier, the influence of agricultural cooperatives on the LDP policymaking declined in the 2000s. The 2000s were a suitable time for manufacturers to announce their intention to practice agriculture (and/or agriculture-related businesses).

Jun-ichiro Koizumi, a decisive politician who served as the Prime Minister and President of the LDP from 2001 to 2006, created a new framework to reflect *Keidanren* and *Doyukai*'s opinions in agricultural policymaking.

Koizumi established advisory councils to design policy suggestions, including for agricultural reform. Many of the council members were from *Keidanren* and *Doyukai*. Through these councils, *Keidanren* and *Doyukai* increased their influence on the LDP's policymaking.

However, it should be noted that although the LDP supports the concept of aggressive agriculture, manufacturers entering the agricultural sector face two major problems. First, they face difficulty in obtaining laborers with high farming skills. The protection of crops and/or livestock from physiological disorders is a critical problem in farming. To minimize the risk of such disorders, farmers train themselves by carefully observing the growth processes of crops and livestock and by dedicating care and attention to them. This kind of training is not compatible with factory-style labor management. In addition, skilled farmers tend to adjust their tasks based on the unpredictable natural conditions of farms and often prefer not to follow the inorganic time framework determined by the clock.

Obtaining agricultural land is another problem that manufacturers face when beginning agricultural activities. Agricultural landowners prefer selling or renting agricultural land to their friends and relatives. This implies that manufacturers may need to acquire strong human networks in rural communities before joining the agricultural sector.

As discussed in Section "Industry 4.0 and Japan's agricultural land-use policy reform", Industry 4.0 may not solve these two problems.

Inaccuracy of Cropland Ledgers

When implementing the ALR, each municipality office under the ALC compiled special ledgers that recorded the location, owner's name, and user's name for each parcel of agricultural land in the municipality. These lodgers were (and remain) labelled "cropland ledgers (CLs)."

When the ALC was reorganized into the AC in 1951, the management of CLs was transferred from the ALC to the AC. The AC is expected to update its CLs record whenever there are changes to any parcel of agricultural land. However, it often fails to do so, resulting in widespread inaccuracies within CLs.

Municipal governments are responsible for implementing the MAFF's agricultural policies at the local level. Thus, if the information in CLs is incorrect, it becomes difficult for local governments to carry out this responsibility.

Previously, when agricultural cooperatives could efficiently organize all farmers in their jurisdictions, they had accurate information about the existing state of agricultural land use in their constituencies. Thus, local governments implemented agricultural policies to maintain close ties with agricultural cooperatives. However, these cooperatives have already lost their ability to organize local farmers. Therefore, establishing a new agricultural land-use system is necessary.

In 2014, the MAFF requested that municipalities digitalize CL information and make it accessible to the public via the Internet. In doing so, however, the MAFF did not notice any inaccuracies in the CLs. This implies that the MAFF missed the opportunity to comprehensively reform the CL system.

Without accurate information about the agricultural land-use situation, it is difficult for municipal governments to implement the MAFF policies. A typical example is the Program for Direct Payments for Efficient Use of Paddy Fields (PDPEUPF). Under the PDPEUPF, farmers receive subsidies if they grow non-rice crops (designated by the MAFF). In the PEPEUPF, municipal governments are responsible for surveying the types of crops grown on every parcel of agricultural land in their jurisdictions. However, because the information about CLs is inaccurate, it is almost impossible for municipal governments to conduct exhaustive surveys. Consequently, inappropriate applications of PDPEUPF subsidies allegedly tend to occur. The Board of Audit of Japan (BAJ) highlighted this issue in 2022.[7] However, the MAFF did not present any solution.

Without accurate information about the existing agricultural land-use situation, introducing participatory democracy would be difficult. In such a situation, heavy-handed politics (and strengthening discrete administrative bodies, such as the central and municipal governments) may be the only solution for preventing disorders in agricultural land use. In fact, the

---

[7] https://report.jbaudit.go.jp/org/r04/2022-r04-0289-0.htm.

MAFF has opted to introduce a heavy-handed system for agricultural land use, as discussed in Section "Industry 4.0 and Japan's agricultural land-use policy reform".

Idle Agricultural Land

Idle agricultural land (unused for agricultural or non-agricultural purposes) is increasing nationwide. According to an estimate made by the MAFF, the total acreage of idle agricultural land increased from 131 thousand hectares in 1975 to 425 thousand hectares in 2015.[8] Idle agricultural land causes various problems in neighboring agricultural land. Since water does not flow smoothly across parcels of idle paddy fields, neighboring rice farmers often face difficulties when irrigating and/or draining their fields. Additionally, unused agricultural land frequently becomes a breeding ground for harmful insects and pathogenic bacteria.

The following two patterns regarding how agricultural land becomes idle are often observed.

Pattern A: Provisional registration
The ALA allows non-farmers to purchase a parcel of agricultural land only if the AC permits its conversion for non-agricultural purposes. A special action called provisional registration creates a loophole in this regulation. For instance, Person A, who wants to purchase agricultural land for non-agricultural purposes, and Person B, an agricultural land owner, visit the Regional Legal Affairs Bureau together and submit a contract requesting the transfer of ownership of a certain parcel of agricultural land from Person B to Person A. This contract is called "provisional registration." Usually, as the contract for provisional registration is submitted, Person B quits farming, and Person A pays money for the parcel of agricultural land. However, the AC may not approve the agricultural land conversion plan, and in this case, the land becomes idle.

---

[8]According to the agricultural censuses, the total acreage of agricultural land was 4.4 million hectares in 2020.

Pattern B: No appropriate successors after the farmers' death
The ALA stipulates that any individual planning to purchase a parcel of agricultural land must receive the AC's approval by demonstrating sufficient farming ability. However, this does not apply to the inheritance of agricultural land. For example, a farmer's successor with no ability or will to pursue farming can own agricultural land following the farmer's death. If the successor retains ownership of a parcel of agricultural land with the expectation of profiting from its future sale for non-agricultural purposes, they may be reluctant to lease it to a farmer (once the land is rented to a tenant farmer, the successor may be required to pay compensation money to the tenant if the land is later sold for non-agricultural purposes). Consequently, the parcel of agricultural land becomes idle.

Non-farmers' succession of agricultural land presents another serious problem: regular membership in agricultural cooperatives. Reflecting the rural patriarchal culture, regular membership in an agricultural cooperative is usually transferred from the previous household head to a new one. However, if the new head does not engage in farming, agricultural cooperatives must not provide regular membership to the new head. However, these cooperatives often do not seriously investigate whether the new head satisfies the conditions for regular membership. In addition, because agricultural cooperatives' major revenue comes from banking activities, they are motivated to maintain the new head as a regular member. Consequently, the total number of non-farm households that hold memberships in agricultural cooperatives is now almost equivalent to that of farm households (Table 4). This means that agricultural cooperatives have already lost their identity as agricultural organizations.

Cropland Intermediary Management Program

Previously, rural Japanese communities were characterized by numerous small farmers, and agricultural land use within a community was well organized under the leadership of agricultural cooperatives. However, these cooperatives have become incapable of organizing farmers. Therefore, building a new system for agricultural land-use planning is an urgent issue in the Japanese agricultural sector. The MAFF employs an authentic approach to agricultural land-use problems, as shown further in the chapter.

In 2014, the MAFF launched a new system called the Cropland Intermediary Management Program (CIMP), which arranges contracts

Table 4. Number of households that hold regular/associate membership of agricultural cooperatives.

(in thousand)

	Number of households which hold regular membership of agricultural cooperatives (1)	Number of farm households (2)	Number of non-farm households which hold regular membership of agricultural cooperatives (3) = (1) − (2)	Number of households which hold associate membership of agricultural cooperatives (4)
1990	4,859	3,835	1,024	2,598
1995	4,780	3,444	1,337	2,972
2000	4,574	3,120	1,454	3,163
2005	4,350	2,848	1,502	3,444
2010	4,068	2,528	1,540	4,061
2015	3,771	2,155	1,616	4,814
2020	3,462	1,747	1,715	5,108

Source: Ministry of Agriculture, Forestry and Fisheries, Government of Japan, *Statistics on Agricultural Cooperatives, Agricultural Census.*

among those who want to sell and/or lease agricultural land and those who want to purchase and/or rent agricultural land. As a special entity for enacting the CIMP, the Cropland Intermediary Management Institute (CIMI) was established in each prefecture, with the governor designating a public interest-incorporated foundation as a CIMI after receiving approval from the prefectural assembly.

Under the CIMP, landowners and farmers use the CIMI carte blanche to decide who should purchase and/or rent agricultural land. In this sense, the CIMI has strong administrative power as a dictator. The CIMP stems from the idea that top-down decision-making is efficient for agricultural land-use planning.

In 2015, the government abolished the election system for AC members. This implies that all members of the AC should be appointed by mayors (without formal elections).

In 2025, the government abolished the PPAMFR's framework of leasehold contracts. This means that most leasehold contracts for agricultural land must be under the control of the CIMI (as mentioned earlier, leasehold contracts under the ALA are limited). This new policy indicates the development of authoritarianism in Japanese agricultural land use.

Industry 4.0 and Japan's Agricultural Land-Use Policy Reform

There is a view that Industry 4.0 will support the CIMP's new agricultural land-use system. Unlike mechanization before Industry 4.0, artificial intelligence (AI), combined with information technology (IT), enables flexible thinking. In fact, the MAFF has high expectations for AI and IT development. The MAFF presented a new idea, called "smart agriculture," suggesting that farming should be dramatically automated (including unmanned operation) based on AI and IT. In 2013, the MAFF created a project team to promote smart agriculture. Based on discussions with the team, it prepares various subsidies when farmers purchase high-technology agricultural machinery and facilities equipped with AI and IT. Currently, smart agriculture is not sufficiently profitable to operate without subsidies from the MAFF. However, accumulating research and development (R&D) in smart agriculture may improve profitability in the near future. Subsequently, a perfectly automated (i.e., unmanned) operation system can be established. Instead of traditional small-scale farmers, modern companies (such as manufacturers) will bear the task of introducing such systems into farm management.

AI will also support the CIDI's decision-making regarding which farmers should farm which parcel of agricultural land. As the CIDI's decisions significantly influence the rural economy, CIDI officials face substantial pressure from rural families. AI support in the local economy will provide great relief to CIDI officials.

From terrestrial analyses of land use, Godo proceeds on to analyze the impact of Industry 4.0 on aquaculture to cover both terrestrial and aqua-based perspectives.

Industrial Revolution 4.0 and Japan's Fishery System

Microeconomics textbooks often refer to the fishery industry as a typical example of a situation in which *laissez-faire* practices may lead to undesirable outcomes. If all people were allowed to fish freely, they would attempt to catch as many fish as possible, and fishery resources would be irreversibly depleted. This is commonly known as the "tragedy of the commons." Conversely, if all people with access to fishery resources agreed to maintain the total fish catch at a renewable level, fishing could be sustainable indefinitely. Thus, organizing all fishermen according to their common interests is an essential problem in the fishery industry.

Japan has a long history of a unique system called *Isson Sen-you Gyojo Seido*, for the sustainable use of fishery resources. However, *Isson Sen-you Gyojo Seido* is currently in decline. This implies that Japan is currently facing a serious risk of the tragedy of the commons. Can the Industrial Revolution 4.0 help Japan to develop a new fishery system in place of *Isson Sen-you Gyojo Seido*? By answering this question, this study discusses the problems and potential of the Japanese fishery industry.

Japan's Geographical Characteristics from the Viewpoint of the Fishery Industry

Before examining Japan's fishery policy, it is useful to review Japan's geographical characteristics from the viewpoint of the fishery industry.

Japan is an island country located at the eastern end of the Eurasian Continent. Most of Japan lies between 30° and 40°N latitude. Japan's western and eastern coasts face the Sea of Japan and the Pacific Ocean, respectively. Both sides have abundant fishery resources through different mechanisms.

The Japan Current, the world's largest warm current, and the Kuril Current, the world's largest cold current, converge along Japan's eastern coastal and offshore areas. This convergence creates a high biological density and rich diversity of aquatic animals on Japan's eastern side.

Japan's western side faces the Sea of Japan. The warm Tsushima Current enters the Sea of Japan through the Tsushima Strait and flows to the northern end of the sea (little ocean water enters the Sea of Japan through other straits, other than the Tsushima Strait, which are very narrow and cold currents north of the Sea of Japan with weak pushing power). Consequently, the water temperature of the Sea of Japan is higher than that of other sea areas at the same latitude worldwide.

In addition, rivers on the Eurasian Continent (e.g., the Amur River) introduce cold, fresh water rich in oxygen. The Sea of Japan's combination of high dissolved oxygen levels and high temperature creates a favorable environment for many types of aquatic life. Consequently, although fish species differ from those in the Pacific Ocean, the Sea of Japan has abundant fishery resources.

Various fishing methods have been developed at coastal and offshore areas of Japan. Many of them are "bycatch style." For example, in stationary net fishing, one of the most popular fishing methods in Japan, fishermen catch various fish species. Fishing results significantly fluctuate daily according to changing water and weather conditions.

In such situations, protecting fishery resources from overfishing is difficult. One possible regulation is to set the maximum total catch for each fish species. However, this is difficult to achieve in bycatch-style fishing. Assuming that the total catch of a certain species approaches its maximum amount, then fishermen abandon the caught fish to continue fishing for other species. This results in the wasteful use of fishery resources. Thus, bycatch-style fishing requires a sophisticated system for the joint use of fishery resources. In addition, because various fishing methods are available in a fishing area at coastal and offshore areas, the allocation of fishing methods is a difficult problem.

The History of Japan's Fishery System Before the Pacific War

Many of the current regulations and rules in Japan's fishing industry are rooted in the traditional feudal-era system. Thus, it is useful to briefly review the history of Japan's fishing system before the Meiji Restoration of 1868, which is commonly recognized as the beginning of Japan's modernization.

During the feudal era, each fishing village had its own (usually unwritten) rules governing the use of fishery resources. These rules specified how to protect fishery resources (e.g., setting no-fishing periods and/or areas) and how to allocate fishing efforts and benefits from common fisheries (e.g., beach seine fishery). These rules evolved naturally based on historical experiences and legends. In principle, only those who lived in fishing villages were allowed to fish. This system was called *Isson Sen-you Gyojo Seido* and represented the villagers' exclusive usage rights over fishery resources as vested by the feudal lords. Fishermen in each fishing village form an autonomous society. The chief fisherman, called *Nanushi,* organized all the fishermen in the village and paid taxes to the feudal lord.

During the Meiji Restoration, the Meiji government believed that malpractices of the feudal system, such as the *Isson Sen-you Gyojo Seido,* should be eradicated. In 1874, the Meiji government launched a plan for a new framework for its fishery policy called *Kaimen Kan-yu Shakku Seido,* a leasehold system for the sea surface established between the government and the fishermen. In *Kaimen Kan-yu Shakku Seido,* the Meiji government acquired ownership of the water surface, and fishermen were granted fishing rights upon paying rent to the Meiji government. Under this plan, the Meiji government abolished *Isson Sen-you Gyojo Seido.* However, it failed to implement an adequate surveillance system for fishing activities under *Kaimen Kan-yu Shakku Seido.* Consequently, unregulated practices, such as illegal fishing and overfishing, became rampant following the introduction of *Kaimen Kan-yu Shakku Seido.* In 1875, the Meiji government canceled *Kaimen Kan-yu Shakku Seido* and allowed fishermen to observe *Isson Sen-you Gyojo Seido* as they had done previously. However, *Isson Sen-you Gyojo Seido* was not subject to modern laws. Since the Meiji government raised the slogan of "Quit Asia and Join Europe," it sought to formalize *Isson Sen-you Gyojo Seido* within the construct of modern law. In fact, it became the Fisheries Act, which was established in 1903 (to distinguish the Fisheries Act after the Pacific War, this paper hereafter refers to the Fisheries Act in the pre-Pacific War period as the "Meiji Fisheries Act"). The Meiji Fisheries Act stipulated that each fishing village should have its own fishermen's union to organize communal fishing among all the fishermen in the village. In fact, the fishermen's unions adopted most of the rules of *Isson Sen-you Gyojo Seido.*

Fishermen's unions were established as self-governance bodies and were prohibited from engaging in economic activities such as marketing and banking. However, with the economic development in fishing villages, fishermen's unions sought to provide banking services to fishermen. In 1933, fishermen's unions were reformed into fishermen's cooperatives, which succeeded the unions' function as organizers of *Isson Sen-yu Gyojo Seido*, and they were allowed to engage in economic activities on their own behalf. In 1943, as Japan became part of the mobilization of the entire country, fishermen's cooperatives were reformed into fishermen's associations. The government controlled the entire process of marine products marketing. All marine products were collected by fishermen's associations and shipped to wholesalers in compliance with government orders.

After the Meiji Restoration, owing to the inflow of modern technologies from European and North American countries, new types of fishery technologies emerged, such as fixed-net fishery and demarcated fishery, which did not exist under *Isson Sen-yu Gyojo Seido*. New fishery technologies offered higher-fishing efficiency but increased the risk of overfishing. In addition, conflicts arose between those who wanted to continue conventional fishing under *Isson Senyu Gyojo Seido* and those who wanted to introduce new technologies. To solve these problems, the Meiji government introduced licensing systems for newly developed fishery technologies, namely, the fixed-net fishery license system, the demarcated fishery license system for aquaculture, and the special fishing rights system for other types of new fishery technologies. Each applicant was required to specify the water location(s) and type(s) of fishery. The license term was, in principle, 20 years. However, once the license was issued, the licensees were allowed to have their licenses renewed without limitation, as long as they demonstrated the willingness and ability to continue the licensed activity.

Various economic entities, such as fishermen's unions, private companies, and individual citizens, applied for the licenses, but the government's screening process was not transparent. However, previous studies have identified a tendency for larger companies with greater financial resources receive favorable treatment during the screening process.[9] This reflected the fact that the initial costs associated with the new types of fishing were much higher compared to those of traditional practices, and the propertied classes were dominant in Japanese politics during the pre-Pacific War period.

After the 1930s, Japan gradually shifted to a military society, which affected its fishery policy. In 1943, as part of Japan's militarization policy, fishermen's unions were reformed into fishery associations that controlled their daily lives beyond fishing activities. Each fishing village had its own fishery association, in which all fishermen in the village were legally required to participate. All fishery associations in Japan were subordinate to the National Fishery Association, which was controlled by the government.

Fishery Policy Reform in the Early Post Pacific War Period

From 1945 to 1952, Japan was democratized under the leadership of the General Headquarters of the Allied Forces (GHQ). In 1948, as part of the GHQ's democratization policy, fishery associations were replaced by new organizations called fishermen's cooperatives (while fishermen's organizations were also called "fishermen's cooperatives" between 1933 and 1944, the system of fishermen's cooperatives after the Pacific War is different from theirs). Unlike in fishery associations, fishermen were not legally required to participate in cooperatives. However, as discussed in the following paragraphs, some types of fisheries were accessible only to members of the fishermen's cooperatives.

The Meiji Fisheries Act was replaced by an entirely revised version in 1949 as the centerpiece of the GHQ's reform of Japan's fishery policy. Hereafter, this paper refers to the "Postwar Fisheries Act". The basic structure of the Postwar Fisheries Act remained unchanged until 2018 (to distinguish the Fisheries Act before 2018, this paper hereafter refers to today's Fisheries Act as the "2018 Fisheries Act").

As shown in Figure 4, the Postwar Fisheries Act established three administrative categories of fishery types: *Kyoka Gyogyo* ("licensed fishery"), *Gyogyo-ken Gyogyo* ("fishing-right-based fishery"), and *Jiyu Gyogyo* ("free fishery").

---

[9]For example, please refer to Kase (2014).

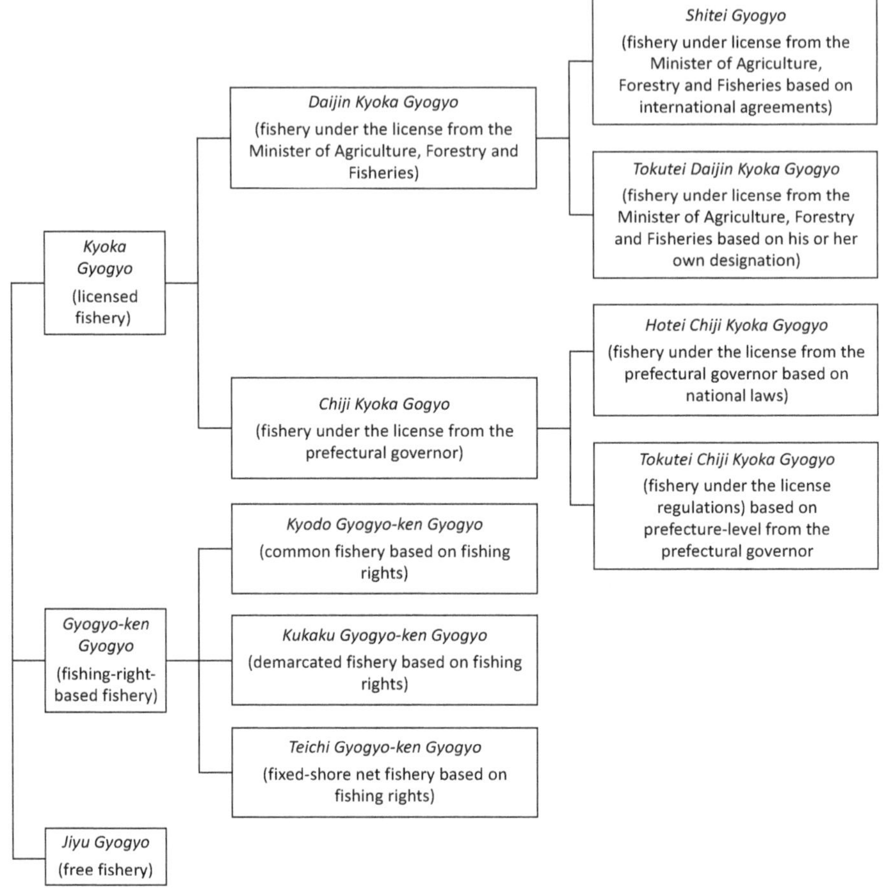

Figure 4.   Classification of fishery under the Postwar Fisheries Act.

Kyoka Gyogyo applies to fishing types that are so effective in catching fish that excessive competition can occur among fishermen if the technologies are used without regulation. In *Kyoka Gyogyo*, only those who receive the licenses from the authorities are allowed to engage in fishing activities. The licensing authority differs by type of fishery: *Daijin Kyoka Gyogyo* or *Chiji Kyoka Gyogyo*. *Daijin Kyoka Gyogyo* is under the auspices of the Minister of Agriculture, Forestry and Fisheries, and *Chiji Kyoka Gyogyo* is under the prefectural governor.[10]

---

[10]The Ministry of Agriculture, Forestry, and Fisheries was named the Ministry of Agriculture and Forestry until 1978. For simplicity, this paper refers to it as the Ministry of Agriculture, Forestry, and Fisheries for the period prior to the renaming.

Daijin Kyoka Gyogyo applies to the types of fishing that the Japanese government is required to control following international negotiations on fishery resources in international waters. There are two types of *Daijin Kyoka Gyogyo: Shitei Gyogyo* ("fishery licensed under the Minister of Agriculture, Forestry and Fisheries based on the international agreements") and *Tokutei Daijin Kyoka Gyogyo* ("fishery licensed under the Minister of Agriculture, Forestry and Fisheries based on their own designation"). In the former case, the Minister of Agriculture, Forestry and Fisheries first establishes national quotas for the total number and tonnage of ships. It then allocates quotas to applicants with sufficient fishing abilities. In the latter case, the Minister of Agriculture, Forestry, and Fisheries issued licenses to those with appropriate fishing plans and ships without setting quotas at the national level.

There are two types of *Chiji Kyoka Gyogyo: Hotei Chiji Kyoka Gyogyo* ("fishery licensed under the prefectural governor based on the national laws") and *Tokutei Chiji Kyoka Gyogyo* ("fishery licensed under the prefectural governor based on the prefecture-level rules of regulation"). In *Hotei Chiji Kyoka Gyogyo*, to protect marine resources at the national level, the Minister of Agriculture, Forestry and Fisheries determines each prefecture's quota based on the total number and tonnage of ships. The governor then allocates quotas to owners and ships in their territorial waters. In *Tokutei Chiji Kyoka Gyogyo*, the prefecture governor has the authority to introduce licensing systems for certain types of fishing methods to protect fishery resources in their territorial waters based on their own judgment. *Tokutei Kyoka Chiji Gyogyo* covers many types of fishing methods, and the list and content differ according to prefecture.

Gyogyo-ken Gyogyo is Japan's unique system for coastal fisheries, whereby members of fishermen's cooperatives are favorably treated to obtain fishing rights. There are three types of *Gyogyo-ken Gyogyo: Kyodo Gyogyo-ken Gyogyo* for small-scale fixed-shore net fisheries and types of fishing performed under *Isson Sen-yu Gyojo Seido* in the pre-Pacific War period; *Kukaku Gyogyo-ken Gyogyo* for demarcated fisheries; and *Teichi Gyogyo-ken Gyogyo* for large-scale shore net fisheries.

In *Kyodo Gyogyo-ken Gyogyo*, only members of the fishermen's cooperatives were allowed to engage in fishing, as organized under the cooperatives. This implies that *Kyodo Gyogyo-ken Gyogyo* can be viewed as a reformed version of *Isson Sen-yu Gyojo Seido*. For *Kukaku Gyogyo-ken Gyogyo* and *Teichi Gyogyo-ken Gyogyo*, any economic entity (whether natural or legal seeking to engage in demarcated fishery and large-size shore net fishery should apply to the prefectural governor for fishing rights. If multiple applicants want to engage in the same fishing type and the same location, the Postwar Fisheries Act stipulates that the right should be granted in the following priority order.

- The first (highest) priority is the fishermen's cooperative in the fishing village.
- The second priority is the legal entity formed by more than 70% of the individual fishermen in the fishing village.
- The third priority is legal or natural people engaged in fishing in fishing villages or elsewhere.
- The fourth (and lowest) priority is others.

As discussed previously, in pre-war Japanese licensing systems for demarcated fisheries and fixed-shore net fisheries, wealthy classes living in urban areas tended to receive preferential treatment in the government's screening of applicants. The GHQ considered such a treatment undemo-cratic. This was the reason for listing fishermen's cooperatives at the top of the priority order.

The terms for fishing rights granted under *Kyodo Gyogyo-ken Gyogyo*, *Kukaku Gyogyo-ken Gyogyo*, and *Teichi Gyogyo-ken Gyogyo* are 10 or 5 years. However, as long as those who hold the rights demonstrate their ability and willingness to continue engaging in the fishery, their rights will be renewed.

Jiyu Gyogyo refers to all types of fishing that are not included in the lists for *Kyoka Gyogyo* and *Gyogyo-ken Gyogyo*. Anyone is allowed to engage in *Jiyu Gyogyo* freely. Pole-and-line fishing for sea bream is an example of *Jiyu Gyogyo* activity.

Sea-Area Fisheries Adjustment Committees

The Postwar Fisheries Act stipulates the establishment of new administrative committees for the Japanese government's fishery policy, called the Sea-area Fisheries Adjustment Committees (SFACs) and Wide Sea-area Fisheries Adjustment Committees (WSFACs).

Before providing further information on SFACs and WSFACs, it would be useful to review the relationship between Japan's sea area system and local governance. Japan has 66 marine zones as its basic fishery unit. Each marine zone belongs to one of the 47 prefectures (Hokkaido has 10; Nagasaki has 4; Fukushima and Kagoshima have 3 marine zones; Aomori, Ibaragi, Tokyo, Niigata, Hyogo, Shimane, Yamaguchi, Sag, and Kumamoto each have 2 marine zones; the remaining 33 prefectures each have only 1 marine zone).

The Postwar Fisheries Act requires prefectural governors to establish an SFAC for each marine zone.[11] Each SFAC comprises 15 members: 9 are elected fishermen, and the governor appoints the remaining 6 (4 are academics and the remaining 2 are representatives of the public interest). The governor is required to obtain approval from the corresponding SFAC when they want to make an important decision, such as issuing fishing rights in a marine zone within their constituency.

The 66 marine zones are grouped into three wide-sea areas: the Pacific Ocean Wide Sea area, the Sea of Japan/Kyushu West Wide Sea area, and the Seto Inland Sea Wide Sea area. Each wide sea area has its own WSFAC. The WSFAC comprises representatives of the corresponding marine zones, fishermen, and academics. The Minister of Agriculture, Forestry and Fisheries is required to obtain approval from the corresponding WFAC when they seek to make important decisions, such as establishing *Daijin Kyoka Gyogyo*.

The establishment of the SFACs and WFACs reflected the GHQ's belief that the introduction of democratic administrative committees was necessary to democratize Japan's fishing policy.

Problems with the Postwar Fisheries Act

Apparently, the Postwar Fisheries Act attempted to solve difficult problems such as preventing the tragedy of the commons and mediating conflicts among fishermen. However, the Postwar Fisheries Act (as well as the 2018 Fisheries Act as discussed later) has the following three problems:

(i) **Unpreparedness for the declining organizational ability of fishermen's cooperatives:** The system of fishermen's cooperatives is the embodiment of *Isson Sen-yu Gyojo Seido*, and the Postwar Fisheries Act implicitly assumes that fishermen's cooperatives can organize

---

[11] If necessary, governors in neighboring prefectures use additional administrative committees, called the United Sea-area Fisheries Adjustment Committees, to discuss marine use problems within neighboring marine zones.

fishermen within their jurisdictions. This assumption is correct when fishermen perceive a strong traditional sense of unity. However, with the development of individualism in the post-Pacific War period, traditional culture and practices in fishing communities have gradually faded. Hence, it is becoming difficult for fishermen's cooperatives to organize fishermen. The Postwar Fisheries Act is not prepared for this situation.

(ii) **Shortage of outsiders in the fishing industry:** Work ethics among Japanese laborers are changing; younger generations are moving away from manual work, including fishing. Moreover, the birthrate is falling and, owing to the changing lifestyles of Japanese families, younger generations are less interested in continuing their family businesses. Hence, many fishermen's cooperatives are suffering from population aging and a shortage of successors. This problem can be resolved by recruiting new fishing community members. However, cooperatives fear that new members will not honor the community's traditions; hence, they are reluctant to recruit them.

Some outsiders have shown interest in starting demarcated fisheries. However, it is difficult for outsiders to obtain fishing rights because the Postwar Fisheries Act prioritizes fishermen's cooperatives. The (self-proclaimed) reform-minded groups claim that fishermen's cooperatives have become "old guards" to protect the interests assured by the Postwar Fisheries Act. However, these reform-minded groups do not suggest who should replace the fishermen's cooperative roles in avoiding and/or mediating conflict among fishermen in limited coastal areas. Meanwhile, conservative groups point out that if the organization of fishermen's cooperatives weakens further, the risk of disorder in the allocation of fishery resources among fishermen and the tragedy of the commons will increase. Conservative groups suggest that national and prefectural governments provide more support to fishermen's cooperatives to avoid and/or mediate conflicts among fishermen in limited coastal areas. However, this argument does not detract from the bitter reality that the organizational ability of fishermen's cooperatives is fading.

(iii) **Inability to prevent overfishing during the emergence of new fishing technologies:** In *Kyoka Gyogyo*, the Minister of Agriculture, Forestry, and Fisheries and the prefectural governors capped the number and tonnage of ships. This type of regulation is called "input control." Conversely, major developed countries in Australasia, Europe, and North America employ "output control-type regulations in which governments set a national-level total allowable catch (TAC) for each type of fish based on scientific evidence. Subsequently, they allocated it to each fisherman (including a legal person) and/or ship as either an individual quota (IQ) or an individual transferable quota (ITQ).

The post-Pacific War period witnessed the emergence of various new fishing technologies that enabled Japanese fishers to catch more fish at the same input level. This has rendered input control regulations ineffective in preventing overfishing.

In 1997, the Japanese government introduced a TAC system for eight fish species: Alaska pollack, bluefin tuna, Japanese common squid, Japanese jack mackerel, Japanese sardine, mackerel, Pacific saury, and snow crab. However, considering that other major advanced countries employ TAC for at least dozens of fish types, Japan's attitude toward output control seems halfhearted. Experts have asserted that the Japanese government is unable to provide sufficient scientific evidence to support the establishment of TACs. Therefore, Japan's TACs are too high to protect fishery resources.[12] To worsen the situation, penalties on fishermen whose total fish catch exceeds the TAC are not strong enough to enforce the TAC.[13]

Miyagi Prefectural Governor Murai's Attempt to Deregulate Fishing Rights

Over time, the significance of the above-mentioned problems has continued to grow. However, there was no major amendment to the Postwar Fisheries Act until 2018. Miyagi Prefectural Governor Murai raised important criticisms of the Postwar Fisheries Act in 2011. This section reviews Murai's attempts and considers their implications.

Miyagi Prefecture is known as one of the major aquaculture areas in Japan. Oyster farming in this area is highly profitable. Although suitable

---

[12] For example, see Katsukawa (2012).

[13] Those who violate TAC regulations are fined lightly by the government. Further details can be found in Katsukawa (2012).

locations for oyster farming rafts are geographically limited, many have expressed an interest in entering or expanding oyster farming production. As described in Section "Fishery policy reform in the early post Pacific War period", the Postwar Fisheries Act provides exclusively favorable treatment to fishermen's cooperatives in setting *Kukaku Gyogyo-ken*. Fishermen's cooperatives decide which fishermen should do which parts of the fishing areas based on conventions accepted by existing fishermen's communities. Other than the members of fishermen's cooperatives, it is difficult to engage in oyster farming.

This frustrated those who sought to participate in oyster farming and were not allowed to. For example, in 2007, the Japan Economic Research Board served as the research and consulting arm of the Development Bank of Japan. One of the earliest economic think tanks in postwar Japan claimed that fishermen's cooperatives unfairly monopolized aquaculture benefits and advocated outsider entry.[14]

At this time, the Great East Japan Earthquake occurred, and the subsequent tsunami destroyed oyster farming equipment and facilities (e.g., ladders, boats, and warehouses) in Miyagi Prefecture. However, the natural features of oyster farming locations were not severely damaged. Thus, if the equipment and facilities are reinstalled, oyster farming could return to its original condition. One difficulty is that fishermen's funding is limited, as most run small family businesses. Some politicians, such as Miyagi Prefecture Governor Murai, have proposed that if large private companies were to invest in oyster farming, they could accelerate recovery from tsunami damage.

Two months after the Great East Japan Earthquake, Murai voiced the following idea in TV and newspaper interviews: To promote the reconstruction of Miyagi Prefecture's fishery industry, Miyagi Prefecture should have a Fisheries Reconstruction Special Zone (FSZR) where outsiders are given top priority to receive exclusive aquaculture rights. Murai's idea was supported by market fundamentalists, whereas conservatives opposed it. The Reconstruction Design Council (RDC), an advisory board for the Prime Minister, was pressured to respond in one way or another to Murai's idea. One month after Murai's announcement, the RDC put forth a proposal, "the first recommendations for the FSZR," suggesting that legal persons (mainly established local fishermen) should be given the same priority as those in the fishermen's cooperative in receiving exclusive aquaculture rights. Following this proposal, the Japanese government established the FSZR in Miyagi Prefecture in December 2012.

---

[14]See the Japan Economic Research Institute (2007).

The FSZR establishment frustrated fishermen cooperatives nation-wide. In particular, the Miyagi Prefecture Fisheries Cooperative (MPFC), which conducts oyster farming at many water locations in Miyagi Prefecture, has repeatedly criticized Murai's ideas, arguing that the FSZR should be abolished. Moreover, even after the FSZR was established, the entry of outsiders was limited. To date, Momonoura Producer of Oyster Consolidated Company (MPOCC) is the only company that has initiated oyster farming under the FSZR framework. The MPOCC was founded in August 2012 by 15 local fishermen in the Momoura District of Ishinomaki City in Miyagi Prefecture and began operations in September 2013. Unlike ordinary fishermen, MPOCC employees are engaged not only in oyster farming but also in producing oyster-based foods. To strengthen its financial position and broaden its line of business, MPOCC received capital from one of the biggest wholesalers of marine products in Miyagi Prefecture, Sendai Suisan, in September 2012. Miyagi Prefecture also supported MPOCC by providing subsidies of 550 million yen (nearly 5.2 million USD) as part of the company's establishment.

MPOCC receives favorable treatment in the FSZR; thus, it is expected that MPOCC would be somewhat hostile towards the MPFC. However, the MPOCC applied for membership in the MPFC in September 2012, and the MPFC quickly approved the MPOCC's application. Neither the MPOCC nor the MPFC justified these decisions. A possible explanation for this is that both the MPFC and MPOCC attempted to avoid confrontation. In September 2013, the MPOCC applied for exclusive rights to aquaculture in the Momoura District. Since no other entity presented itself as a competition (not even the MPFC), the MPOCC was permitted to begin oyster farming in Momoura District.

Business performance evaluations of MPOCC's oyster farming have differed: some experts consider it "excellent" while others consider it "poor." Indeed, MPOCC oyster production has been increasing, and the company has maintained a black balance. The Miyagi Prefectural Government (2018) concluded that the MPOCC was successful. However, the MPOCC oyster production rate has been lower than average in Miyagi Prefecture, and the company relies heavily on prefectural subsidies. Furthermore, the company experienced mislabeling issues in 2014 and 2015.[15] Consequently, some citizen groups disagreed with the Miyagi

---

[15] For further details, refer to the Union of Members of the Communist Party of the Miyagi Prefectural Assembly (2017).

Prefecture Government (2018) and concluded that the MPOCC's entrance into oyster farming was unsuccessful.[16]

In August 2018 and 2023, the MPOCC's exclusive right to aquaculture in Momoura District was renewed, as no other entity applied for it. Entities other than MPOCC have never been applied to FSZR.

Clearly, a *laissez-faire* policy is inappropriate for use in fishery resources. Unfortunately, throughout the debate on FSZR, no alternative system for controlling fishery resources has been presented. Whether fishermen's cooperatives should prioritize the allocation of *Kukaku Gyogyo-ken* was discussed. The FSZR experience implies the reluctance of Japan's society to establish a new fishery resource system, despite its necessity.

The 2018 Fisheries Act

The bill for amending the Postwar Fisheries Act was passed through the Diet in 2018 (i.e., the 2018 Fisheries Act), and it became effective in 2020. This amendment has three major implications.

(i) **Output control-style regulations for fishery resource protection:** The 2018 Fisheries Act stipulates that by setting the TAC for major types of fish, Japan's fishery-resource regulation policy should be changed from an input to an output control style. However, considering that the Japanese government's TACs are too large to protect fishery resources, it is unclear whether the 2018 Fisheries Act will be effective in preventing overfishing.

(ii) **Revision of priority order for fishing rights:** In the 2018 Fisheries Act, fishermen's cooperatives were prioritized for fishing rights for *Kukaku Gyogyo-ken Gyogyo*. The 2018 Fisheries Act does not have a priority order. This gives the impression that outsiders' new entry into *Kukaku Gyogyo-ken Gyogyo* has become easier. However, the 2018 Fisheries Act stipulates that the fishing rights of current holders should be renewed if they are used appropriately. This means that the advantage held by the fishermen's cooperatives in receiving

---

[16]For example, see Higashi Nihon Daishinsan Fykkyuu Fukko Shien Miyagi Kenmin Senta (2018).

fishing rights for *Kukaku Gyogyo-ken Gyogyo* will be maintained even after the enactment of the 2018 Fisheries Act.

(iii) **Replacement of members elected by appointed SFAC members:** As mentioned earlier, in the current Fisheries Act, each SFAC comprises 15 members, 9 of whom are elected from among the fishermen. However, under the New Fisheries Act, the Prefectural Ordinance should determine the 10–20 SFAC members. The prefectural governor appointed all members (the election system was abolished). The 2018 Fisheries Act requires members of the fishery industry to consider outsiders' voices in the prefectural governor's decision-making process. This attempt was to reduce the influence of fishermen's cooperatives. Simultaneously, the 2018 Fisheries Act stipulates that more than half of SFAC members should be fishermen.

Market fundamentalists, who regard fishermen's cooperatives as the old guard of anti-reformists, criticize the 2018 Fisheries Act as insufficient because outsiders cannot form a majority in the SFAC.

However, conservatives also criticize those outside the fishery industry for lacking sufficient knowledge on controlling marine resources and mediating conflicts between fishermen.

In summary, the essence of the 2018 Fisheries Act is the reduction in the role of fishermen's cooperatives in the usage of fishery resources. The 2018 Fisheries Act did not present an alternative framework for using fishery resources. It is safe to conclude that the 2018 Fisheries Act failed to solve the three problems mentioned in Section "Problems with the Postwar Fisheries Act".

The Fishery Agency's Vision of the Industrial Revolution 4.0

Under a slogan of "Smart Fishery," The Fisheries Agency (2024) presents the government's vision on how IoT and AI technologies should be applied to the fishery industry (The Fisheries Agency is an independent organ of the Ministry of Agriculture, Forestry and Fisheries). According to the Fisheries Agency (2024), there are two purposes for the application of new technologies: to make fishery resources sustainable and to stimulate the economic growth of the fishery industry.

For these purposes, the Fisheries Agency (2024) points out the following 12 ways to apply the IoT and AI technologies:

(1) Keeping sample boat records of the inputs and outputs of fishing and water conditions in fishing areas.
(2) Collecting information on fish landings at landing ports.
(3) Electric information on the total catch for *Daijin Kyoka Gyogyo.*
(4) Promote the introduction of the IQ system.
(5) Develop a forecast system for water conditions in fishing areas.
(6) Develop a search system for fish shoals using a drone.
(7) Improving the operational system for traditional aquaculture.
(8) Develop a warning system for red tides.
(9) Develop large offshore floating pen-style aquaculture systems.
(10) Develop recirculating aquaculture systems.
(11) Improve the efficiency of marketing systems for marine products.
(12) Protect endangered species of aquatic animals and plants.

However, it is unclear how the IoT and AI technologies should be used. In that sense, "Smart Fishery" can be seen as only an ideal image (or the Fisheries Agency's optimistic image) of the future of the fishery industry.

Proposals for Fishery Policy Reforms

Information and AI technologies can be double-edged swords. These technologies enable fishers to catch more fish in a shorter period. This increases the risk of overfishing in the absence of a protection system for fishery resources.

Theoretically, setting TAC and IQ for each type of fish is desirable. However, it is difficult to determine the TAC and IQ. Fortunately, recent developments in IoT technology have mitigated this difficulty to some extent.

The proposal of this study is as follows:

(1) Reintroduction of an election system for the SFACs
(2) Under the leadership of SFACs, the routines of operations (3), (4), (5), (6), and (7) should be practiced continuously.
(3) Accumulation of data on fishing (when, how much, and what type of fish were caught, and by what fishing method)

(4) Set TAC and IQ based on discussions among scientists, local people, and fish businesspeople.
(5) Continuously update the data on fishing.
(6) Frequently reset TAC and IQ based on updated data.
(7) First, IQ should be given to existing fishermen, and the outcomes, TAC, and IQ should be revised frequently. In each revision, the SFACs should invite applications for new entries into fishing.

Fishermen are exposed to the temptation to underreport their total catch (and sales of marine products). Previously, the complexity of the market channels for marine products was problematic in collecting information on the fishery industry. However, the invoice system, which was introduced in 2022 as part of Japan's tax system reform, makes it difficult to underreport total catches and, therefore, reduces such temptation.

Previously, in Japan's fishery industry, it has been difficult to obtain detailed information on when, where, how much, and what type of fish fishermen catch (partly because "bycatch style" fishing is popular). However, owing to the development of GIS and IoT technologies, it is not technically difficult to collect and accumulate information on who captures what types of marine products in which sea area. It should be noted that trial and error are inevitable when determining appropriate figures for TAC and IQ. Such efforts will foster participatory democracy, which the Japanese society needs to achieve.[17]

Regular Membership Problem of Fishermen's Cooperatives

The Fishery Industry Cooperative Act (FICA) stipulates that only those who engage in fishing for 90 days or more every year are qualified to obtain regular membership in fishermen cooperatives.[18] However, this was not strictly observed.

The government gives fishermen's cooperatives various advantages, such as tax reduction and special treatment under the Antimonopoly Act.

---

[17] Further discussion can be found in Godo (2015).

[18] When the Postwar Fisheries Act was enacted in 1948, those who wanted to obtain regular membership in fishermen's cooperatives were required to fish for at least 30 days per year. This requirement was increased to 90 days in 1962.

Thus, it is unfair for those who do not satisfy the FICA's requirements to hold membership illegally. Previously, it was difficult to determine the number of days each member engaged in fishing. However, based on the development of IoT technology, this investigation is possible. This is also an appropriate way to use Industrial Revolution 4.0.

Fishermen in coastal and offshore areas sometimes receive negative byproducts from the development of the Japanese economy. For example, reclamation and domestic and industrial wastewater may damage fishery resources. Those who inflict damage on the fishery industry may be obliged to pay compensation to those who are damaged. Often, a fishermen's cooperative receives compensation money and distributes it to its membership holders. In some cases, antisocial forces obtain money by holding memberships but do not satisfy the aforementioned 90-day rule.

Conclusion

Japan has rich fishery resources in its coastal and offshore areas. Various fishing methods coexist, and various species of fish are caught in small fishing areas. By developing a hereditary system called *Isson Sen-you Gyojo Seido* in the Tokugawa period, fishermen enjoyed sustainable use of fishery resources. Even after the Meiji Restoration, the framework of *Isson Sen-you Gyojo Seido* was successful in fishermen's unions under the Meiji Fisheries Act of 1903, and in fishermen's cooperatives under the Postwar Fisheries Act of 1948. However, with the development of individualism, *Isson Sen-you Gyojo Seido* has become outdated. Japan requires a new system for fishery resource usage. As discussed in Section "Proposals for fishery policy reforms", Industrial Revolution 4.0 may be useful.

The next section of the volume will move on to the final case study of China.

Bibliography

Fisheries Agency, *Sumato Gyogyo ni Tsuite* (Vision of Smart Fisheries), 2024.

Godo, Yoshihisa, "The Failure of Land-use Planning in Japan", In Murat Yulek (ed) *Economic Planning and Industrial Policy in the Globalizing Economy*, 2015.

Higashi Nihon Daishinsan Fykkyuu Fukko Shien Miyagi Kenmin Senta, *Miyagi-ken Suisantokku no Kensho Kekkano Happyo nitaisuru Watashitachi no Kenkai to Teigen* (Our view on Miyagi Prefectural Government's evaluation on proposal Fisheries Special Zone for Reconstruction), 2018.

Japan Economic Research Institute, *Gyoshoku wo Mamoru Suisangyo no Senryakuteki na Bapponnkaikaku wo Isoge* (A proposal for Strategic Approach to Reform Japan's Fishery Industry and Japan's Seafood Culture), 2007.

Kase, Kazutoshi, *San Jikan de Wakaru Gyogyoken* (A Three-Hour Lecture on the Fishing Right System), Tsukuba Shobo, 2014.

Katsukawa, Toshio, *Gyogyo toiu Nihon no Mondai* (Problems in Japan's Fishery Industry), NTT Shuppan, 2012.

Miyagi Prefectural Government, *Miyagi-ken Ishinomaki-shi Momoura Chiku niokeru Fukko Suishin Keikaku no Kensho* (Review on Reconstruction Program at Momoura District in Ishinomaki City in Miyagi Prefecture), 2018.

Union of Members of the Communist Party of the Miyagi Prefectural Assembly, *Momoura Kaki Seisansha Godo Kaisha niyoru Tachikusan Kaki Ryu-yo Mondai no Tettei Chosa wo Motomeru Moushi-ire* (Proposal for Comprehensive Survey on the Mislabeing Problem of Momonoura Producer of Oyster Consolidated Company), 2017.

Chapter 3

History of Technological Use in Chinese Agriculture and Food Sectors

Tai Wei Lim

Abstract

This chapter on China briefly surveys the history of technological use in Chinese agriculture and food sectors. It adopts area studies and political economic angles to examine this topical matter with a focus on policy analysis related to the utilization of Industry 4.0 technologies.

Introduction and World Historical Background

Approximately 12,000 years ago, humans were completely nomadic peoples, hunting and foraging in lands with all kinds of terrains globally, but global warming soon made cultivation possible as the weather transitioned from the ice age to a warmer and drier climate.[1] As the weather became hotter and drier, nomadic hunters hunting and foraging in elevated lands discovered that plants and animals around them were dying, the surviving animals in the ecosystem migrated to the river valleys in search

---

[1] Harford, Tim, "How the plough made the modern economy possible" dated 27 November 2017 in *BBC World Service* [downloaded on 27 Nov 2017], available at https://www.bbc.com/news/business-41903076.

of water, and their human predators followed accordingly.[2] The world historical transition from nomadic hunter–gatherer to agricultural settler occurred more than 11,000 years ago in Western Eurasia, almost 10,000 years ago in India and China, and more than 8,000 years ago in Mesoamerica and the Andes.[3]

Other sources opine that China has been a nation of peasants for four millennia.[4] As with many ancient civilizations, the Chinese agricultural engineering subsystem began in the Neolithic period with tools and equipment made of stone, wood, shell, and pottery to excavate and relocate topsoil for plant cultivation, and these contraptions make up the majority (97%) of the 102 technologies present then.[5] The Neolithic period began in China approximately 10,000 BC and ended with the emergence of metallurgy approximately 8,000 years later, and the era featured settled communities along river tributary systems, mainly along the Yellow (central and northern China) and Yangzi (southern and eastern China) rivers, and they depended mainly on farming and domesticated animals instead of hunting and gathering.[6]

Therefore, even before the dynastic-era technological revolution, predynastic primitive technologies were already available in China. These primitive technologies evolved in sophistication until the late imperial eras of the Ming (1368–1644) and Qing (1644–1911) dynasties. In the pre-modern era, Chinese knowledge in agriculture rivaled and, in some ways, was better than the West when the Jesuits arrived at the Ming court as early visitors were impressed by highly productive Chinese farming, crop rotation systems, advanced water-raising devices/equipment, and the highly rationalized and industriousness of the Chinese farming industry.[7]

---

[2] *Ibid.*

[3] *Ibid.*

[4] Talal Abu-Ghazaleh Global (TAG, Global), "History of Agriculture in China" undated in TAG Confucius Institute (TAG-Confucius) website [downloaded on 1 January 2023], available at https://tagconfucius.com/article/view/History-of-Agriculture-in-China%C2%A0.

[5] Wu, Shuanglei, Yongping Wei, Brian Head, Yan Zhao and Scott Hanna, "The development of ancient Chinese agricultural and water technology from 8000 BC to 1911 AD" dated 9 July 2019 in Springer Nature [downloaded on 9 July 2019], available at https://www.nature.com/articles/s41599-019-0282-1, p. 7.

[6] Department of Asian Art The Metropolitan Museum of Art, "Neolithic Period in China" dated October 2024 in Heilbrunn Timeline of Art History New York The Metropolitan Museum of Art 2000– [downloaded on 30 October 2004], available at http://www.met museum.org/toah/hd/cneo/hd_cneo.htm.

[7] Talal Abu-Ghazaleh Global (TAG, Global), *Op.cit.*

Even today, innovation is seen as the most important engine of sustainable growth of agricultural productivity in the People's Republic of China (PRC), and it involves technological progress that has the potential to decouple agricultural development from material inputs.[8] This has been the age-old objective and remains very much the case.

Genesis: The Yellow River Civilization

The Yellow River is often known as the cradle of Chinese civilization, where one could find primordial technological tools made of stone, bone and wood for subsistent slash-and-burn cultivation of crops like primitive millets that emerged during the Neolithic period[9] (10,000–2,000 BC[10]). Earlier tools for vegetation cultivation such as lashing (stone axes and knives) and processing harvested crops (stone mills), and early Leisi ploughs made from the late Neolithic copper smelting technique, were developed in the late Neolithic period to the Xia, Shang, Zhou (XSZ) era (2000–771 BC).[11] But due to the restricted supply of copper, majority of the farming tools were made from stone and wood rather than bronze.[12]

In the technological take-off stage from the Xia, Shang, Zhou (XSZ) era (2000–771 BC) to the Chunqiu Zhanguo (CQZG) era (771–221 BC), iron tools and irrigation channels for dryland cultivation emerged, alongside technologies for effective furrowing and dryland cultivation, making crop variations (such as millet and ramie) possible.[13] While these were sporadic innovations, China did undergo massive consolidated agriculture technological innovations.

The inaugural Chinese iron-age revolution in agricultural technology happened when iron agricultural implements were made possible for utilization by early Chinese farmers, e.g., Northern Henan primitive iron

---

[8] Organization for Economic Cooperation and Development (OECD) and OECDiLibrary, "Chapter 6. The agricultural innovation system in China" dated 4 Oct 2018 in OECD Food and Agricultural Reviews, Innovation, Agricultural Productivity and Sustainability in China [downloaded on 1 November 2023], available at https://www.oecd-ilibrary.org/sites/9789264085299-9-en/index.html?itemId=/content/component/9789264085299-9-en [unpaginated online version].

[9] Wu, Shuanglei, Yongping Wei, Brian Head, Yan Zhao and Scott Hanna, *Op.cit.*, p. 11.

[10] Department of Asian Art The Metropolitan Museum of Art, *Op.cit.*

[11] Wu, Shuanglei, Yongping Wei, Brian Head, Yan Zhao and Scott Hanna1, *Op.cit.*, p. 7.

[12] *Ibid.*

[13] *Ibid.*, p. 11.

plough from the Warring States period (475–221 BCE), a flat V-shaped iron piece mounted on wooden blades with handles, and these were smaller-scale tools until beasts of burden-powered ploughs appeared in the 1st century BCE.[14] During the Chunqiu Zhanguo (CQZG) (771–221 BC), Qin Han (QH) (221 BC–220 AD) and Weijin South and North (WJ) (220 AD–581 AD) dynasties, 70–80% of technologies in the Chinese agricultural engineering system were hand-operated with tools that were able to tap into available iron ore resources and shaped into ploughs, shovels, and hoes.[15]

The Yellow River agri-tech development took off during the Qin Han (QH) era (221 BC–220 AD) with shovels and hoes used for furrowing and cultivation of wheat and soy beans (made up 70% overall cultivation), while large-scale irrigation was put in place.[16] During the Qin Han (QH) dynasties (221 BC–220 AD), animal power (e.g., animal-driven traction farming using mostly ox and horses) was utilized in the furrowing function, while some specialized tools were created for rainfall measurements, sowing, soil breakdown, ground leveling, harvesting, and crop processing using smaller hoes, shovels and rakes.[17] The first mouldboard plough emerged in China more than two millennia ago and can be used to slice through a long, thick ribbon of soil and flip it upside down.[18]

The Yangtze River Civilization

Besides the Yellow River civilization in the north, the Yangtze River region became another agri-tech innovation area during the Neolithic Period with tools fashioned out of shell and bone, while rice paddies were developed using slash-and-burn cultivation, and it was in the Chunqiu Zhanguo (CQZG) era (771 BC–221 BC) that ponds and channels for paddy irrigation were started.[19] During the Weijin South and North (WJ) dynasties (220–581 AD) and Sui Tang (ST) dynasties (581–960 AD), irrigation projects proliferated in the Yangtze River area, such as the Lake Tai

---

[14]Nair, Kusum and Gary W. Crawford, "The classical-imperial era" undated in the Britannica [downloaded on 19 November 2023], available at https://www.britannica.com/topic/agriculture/Japan.

[15]Wu, Shuanglei Wu, Yongping Wei, Brian Head, Yan Zhao and Scott Hanna, *Op.cit.*, p. 7.

[16]*Ibid.*, p. 11.

[17]*Ibid.*, p. 7.

[18]Harford, Tim, *Op.cit.*

[19]Wu, Shuanglei, Yongping Wei, Brian Head, Yan Zhao and Scott Hanna, *Op.cit.*, p. 11.

Irrigation Complex in Zhejiang and Jiangsu Province, with its multiple dikes, water ponds, water storage channels, flood diversion, and seawater infusion prevention.[20]

From 1000 to 1800 CE, China had the globe's biggest population and the strongest economy whose population was fed mainly by rice which was only cultivable in the arable lands of the humid rainy southern Chinese provinces and rice became the main source of carbohydrates for all socioeconomic classes[21] (the Confucian hierarchy consists of merchants, artisans, farmers and Mandarins/scholarly gentry, arranged in increasing social status in that order). The gentry officials were perceived as worthy of ascending the hierarchy due to their meritocratic ascension to those positions based on imperial examination results. The farmers were seen as the major producers of food for survival and they needed tools that the artisanal class fashioned and made. The merchants located at the bottom of the social hierarchy were seen as profit-seekers who leeched off others.

Given that rice had emerged as the major source of carbohydrate for all social classes, bigger yields were needed to feed the population. Therefore, there was a need to innovate new technologies to gain more yields. Sedentary farming trumped the declining nomadic lifestyle at this stage of world history because it brought wealth for successful farmers through the accumulation of surplus food stocks in Imperial Rome two millennia ago and 900 years ago in Song China, as the peasants were 5–6 times more productive than nomadic hunter–foragers in sourcing for food.[22]

(Even today, the priority of increasing grain supplies has not changed. The prioritized goal of agricultural innovation in China has been the augmentation of the supply of major agricultural items like grains in the interest of national food security in the PRC.[23] In fact, with technological

---

[20]Wu, Shuanglei, Yongping Wei, Brian Head, Yan Zhao and Scott Hanna, *Op.cit.*, pp. 11–12.

[21]Francesca, Bray, "Rice, Technology, and History: The Case of China" dated Winter 2004 in Education About Asia, Association for Asian Studies (AAS) Education About Asia: Online Archives, Volume 09:3 (Winter 2004): Special Section on Teaching About Asia Through Rice) [downloaded on 1 January 2023], available at https://www.asianstudies. org/publications/eaa/archives/rice-technology-and-history-the-case-of-china/.

[22]Harford, Tim, *Op.cit.*

[23]Organization for Economic Cooperation and Development (OECD) and OECDiLibrary, *Op.cit.*

advancements, China achieved world records in grain output in the 21st century. Some secondary sources/references indicated that the per-ha yield of rice, wheat and corn went up by 9.9%, 44.2% and 28.1% in 2000–2015, respectively, enabling China to attain one of the highest yield levels globally.[24])

During the Weijin South and North (WJ) dynasties (220–581AD) and Sui Tang (ST) dynasties (581–960 AD), technologies like herringbone rakes, ground leveling boards and ploughs for paddy field cultivation and weeding during the ST and Song Yuan (SY) dynasties (960–1380 AD).[25] The Chinese iron plough was innovated with adjustable shovels and sharper heads to enable the furrowing to become more efficient and facilitate operations in large-scale water infrastructure projects.[26] As with most civilizations, then, the plough was managed by men as it was heavy and needed the more muscular males with stronger average upper body strength to operate it, and this reconfigured gender relations in society, while grains like wheat and rice needed more culinary preparation than nuts and berries, so women became domesticated at home to cook food.[27] (Today, such gender imbalances no longer exist in China as agricultural mechanical equipment have been ploughing the crop fields throughout Chinese farmlands like those in Duzhuang Village, Huaxian County, in the central Chinese Henan Province since March 2018.[28])

Technological development from the Qin Han (QH) era (221 BC–220 AD) facilitated furrowing, cultivation, weeding, and crop selection, while the Yangtze River irrigation in locations like Jianhu Pond manages irrigation and flood mitigation for more than 600 hectares of farmland in Zhejiang Province.[29] Subsequent innovations like the three-shared plow, the *louli* (plow-and-sow) implement, and the harrow augmented Chinese agricultural tech capabilities such that, by the end of the Song dynasty in 1279, tech capabilities had matured, although the age-old equipment

---

[24]Organization for Economic Cooperation and Development (OECD) and OECDiLibrary, *Op.cit.*

[25]Wu, Shuanglei, Yongping Wei, Brian Head, Yan Zhao and Scott Hanna, *Op.cit.*, p. 12.

[26]*Ibid.*, p. 7.

[27]Harford, Tim, *Op.cit.*

[28]Xinhua, "Technology reshaping agriculture in China" dated 21 March 2018 edited by Huaxia in Xinhuanet [downloaded on 21 March 2018], available at http://www.xinhuanet.com/english/2018-03/21/c_137054369.htm.

[29]Wu, Shuanglei, Yongping Wei, Brian Head, Yan Zhao and Scott Hanna *Op.cit.*, p. 11.

continued to be found in the Chinese farming sector right up to modernity.[30] Farming innovation during the Song dynasty (960–1279) was nicknamed the "Green Revolution."[31]

Late Imperial China

The Ming dynasty documented many mechanical technologies. Mechanical technologies were visible, such as the water wheel driving a chain pump (illustrated in the 1530 edition of Wang Zhen's *Nongshu*) to drive mills for grinding grain, pounding pulp for paper, forging iron, and pressing oil, and some of them were two-man portable chain pumps (illustrated in Xu Guangqi's *Nongzheng quanshu* or *Complete Treatise on Agricultural Administration* published in 1639).[32]

In the Ming era (1368–1644 AD), the Chinese population doubled in size as new parcels of farmland were worked on by peasants in the Fujian, Guangzhou and Guangxi southern provinces, where new crops were planted and irrigation systems upgraded.[33] In the Ming dynasty, streams were dammed into a canal next to rice fields, and water then ran through them through adjustable wooden sluices. This system was illustrated in the 1530 version of Wang Zhen's *Nongshu (Agricultural Treatise)*, a book completed in 1313.[34] All these irrigation technologies helped the Chinese farming industry expand from river valleys up the mountains and hillsides and down to the river deltas. Generally, humans everywhere started to settle down and colonize fertile but scarcer river valleys, and they farmed by breaking up the topsoil, adding nutrients and making moisture seep deeper into the ground away from sunrays, making it possible for 1/5th of the global population to effectively feed everyone across the globe (at least in the theoretical sense).[35]

(China's population only started to shrink in a controlled and systematic manner after strict population controls were put into place in the contemporary era of the PRC. Providing an increasing population (burgeoning from 552 million in 1950 to 830 million in 1970) with enough

---

[30] Nair, Kusum and Gary W. Crawford, *Op.cit.*

[31] Francesca, Bray, *Op.cit.*

[32] *Ibid.*

[33] Talal Abu-Ghazaleh Global (TAG, Global), *Op.cit.*

[34] Francesca, Bray *Op.cit.*

[35] Harford, Tim, *Op.cit.*

food compelled the state to put in place robust population planning measures, which resulted in total fertility rates dropping from 4.2 in the 1970s to below replacement level and average yearly increases in population growth rate from above 2% in 1950–1970 to 1.8% in 1970–1978, and less than 1% after the late 1990s and in 0.63% in 2001–2004. These demographic restrictions have since been removed in the 21st century gradually due to the advent of a rapidly aging population.[36])

Late imperial China farms stayed small-scale and had basic equipment (some may even say inadequate technology) that squeezed a few more baskets of rice and other grains out of depleted fields, leading to the decline of living standards. Nevertheless, through sheer determination alone, it fed 300 million by 1800 (leading some in the West to label it a 'miracle').[37] The late imperial Qing dynasty continued the tradition of building up storage technologies and facilities. To prevent rebellions, the Qing government set up a meticulous granary system to fend off famines and other disasters like epidemics by dispensing free or low-cost grain.[38]

Up to the modern era, unfenced fields were cultivated by wooden ploughs (sometimes with/without a cast-iron share typically operated with a water buffalo or *shuinui*), harvesting with sickles or billhook cutters fixed with a blade at a hooked point fitted with a handle, shoulder-slung sheaves, and after harvesting, they operated rice-threshers using slats or flails, tossed-winnowing in the wind, then put the rice through hand-pounded mortars or hand-turned mill dehuskers.[39] Meanwhile, the West went through progressive stages of industrial revolutions and developed increasingly sophisticated mechanical technologies to operate in the agricultural fields. At the time when the early phase of the PRC was still using hand-operated tools, tractors had already replaced horse-ploughs in the West, harvesting machines did the jobs in place of entire mobilized local communities, and the activity of hoeing weeds was no longer needed due

---

[36] Huang, Jikun, Jun Yang and Scott Rozelle, "China's rapid economic growth and Its implications for agriculture and food security in China and the rest of the world" undated in the Food and Agriculture Organization (FAO) [downloaded on 1 January 2023], available at https://www.fao.org/3/ag088e/AG088E03.htm.

[37] Francesca, Bray, *Op.cit.*

[38] Talal Abu-Ghazaleh Global (TAG, Global), *Op.cit.*

[39] Nair, Kusum and Gary W. Crawford, *Op.cit.*

to modern chemical herbicides, Thus, capital equipment had taken over human labor in the industrialized West.[40]

Irrigation Technologies

Arid Northern China terrains were not suitable for rice cultivation, so they planted dry-land grains like wheat, millet, and sorghum, but their yields were comparatively low unlike the southern rice paddies that generated surpluses to supply trade and merchant activities nationally.[41] (Even in contemporary China, farmers and producers continue to experiment and implement the planting of high-yield rice and crop rotation.[42] China was one of the earliest to create the hybrid rice, and in the 1980s and 1990s, China's producers were replacing crop varieties in approximately 20–25% of their sown areas during each cropping season, i.e., in approximately every 4–5 years, Chinese farmers completely turned over their technology portfolios.[43])

(The differentiation between Northern and Southern Chinese culti-vated areas continues today. As for contemporary southern China, breed-ers are overriding farmers in developing the agriculturally less arable Guangxi, Yunnan, and Guizhou in the southwestern Chinese mountainous regions. A reason behind the override is the conventional assumption held by officially endorsed plant breeders that farmers are less knowledgeable than breeders.[44] The latter believed that cultivated crops must be geneti-cally uniform and adaptable over varying terrains and that landraces, local domesticated plants developed mostly naturally through adaptation to their natural host environment and open-pollinated varieties indigenous to the southwest, must be replaced by high-yielding varieties in a blanket manner for food security.[45])

---

[40] Francesca, Bray, *Op.cit.*

[41] *Ibid.*

[42] Vernooy, Ronnie, "China Tries Alternative to Industrial Agriculture" dated 20 July 2012 in United Nations University (UNU) [downloaded on 20 July 2012], available at https://ourworld.unu.edu/en/china-tries-alternative-to-industrial-agriculture.

[43] Huang, Jikun, Jun Yang and Scott Rozelle, *Op.cit.*

[44] Vernooy, Ronnie *Op.cit.*

[45] *Ibid.*

Extensive water channels for dryland farming irrigation in northern China were built, e.g., the Zhengguo Channel, Lingzhi Channel and Bai Channel were components of a large irrigation network from the Yellow River during the Qin Han (QH) era (221 BC–220 AD), supplying 40,000 hectares of farmland.[46] Irrigation was very important even for southern China. For more than 2000 years in Jiangnan and Canton (Guangdong), rice paddies with precisely regulated and drained water supply can be constructed by leveling the field and surrounding it with low dykes/bunds with even water depths throughout the paddies (the biggest of which were 20 yards square).[47] An irrigation system emerged in southern China in the Chunqiu Zhanguo (CQZG) era (771–221 BC), centered on water storage (that tapped into Shao Pond), and the Dujiangyan Weir flood control design divides the river into an outer area for flood management and an inner section for irrigating more than 200,000 hectares of cultivated land and rice paddies.[48]

Irrigation was implemented with a wooden, square-paddle chain pump with a foot-stepped radial treadle, and paddies were drained by open ditches/diking, while dung, oil cakes, and ash fertilized the soil, and, with increasing demographic growth, labor-intensive cultivation expanded into the locales with sandy loams, arid hills, and lofty mountains while matching appropriate crops for less fertile land.[49] For example, young rice seedlings need moist soil but rot in standing water, and impounded rainwater in a bunded field may all evaporate before the rice is fully grown, thus peasants in some hilly regions construct tanks to accumulate rainwater released by gravity flow.[50]

Funneling small water streams into hillside terraces, building diversion canals from larger rivers to pump water up into the fields, constructing polder fields (with drainage channels) from land reclaimed from swamps/lakes through the construction of a high solid dyke planted with trees to prevent soil erosion, all help with water management.[51] Irrigation is an ongoing contestation between the Chinese nation and its natural environment. Irrigation remains important even today in contemporary China. Farming

---

[46] Wu, Shuanglei, Yongping Wei, Brian Head, Yan Zhao and Scott Hanna, *Op.cit.*, p. 7.

[47] Francesca, Bray, *Op.cit.*

[48] Wu, Shuanglei, Yongping Wei, Brian Head, Yan Zhao and Scott Hanna, *Op.cit.*, p. 7.

[49] Nair, Kusum and Gary W. Crawford, *Op.cit.*

[50] Francesca, Bray, *Op.cit.*

[51] *Ibid.*

uses approximately 65% of the water in China, and the scarce resource is under pressure from accelerated urbanization and industrialization.[52]

(Even in contemporary China today, irrigation remains a priority for the Chinese agricultural industry. Water insufficiency due to competitive needs of the private sector/manufacturing industry and local household use may mean that there is a limit to sizable increases in the irrigated regions and irrigation output, something crucial to the North China Plain's wheat/maize-growing regions.[53])

Fertilizer Technologies

Some are of the view that China's long history of traditional and ecological farming practices goes back a minimum of 4000 years and it did not utilize chemical fertilizers or pesticides for most of this period, instead relying on traditional methods like using legume crops for nitrogen fixation, crop rotations and intercropping, terracing, diverse crop varieties and recycling human, animal and crop wastes to upkeep soil fertility.[54]

As for the use of fertilizers, a majority of the rice farmers utilized human dung or compost as fertilizers, and by the year 1500, many farmers were acquiring mass-produced lime and soy-bean waste fertilizer, urban human dung generated by urbanites, increasing yearly output to two or three tons per acre in some double-cropping regions.[55] Therefore, some argue that the secret to 4,000 years of land fertility was based on the methodology of "an agriculture without waste," and therefore, it had no utilization of external inputs, featuring small-scale intensive farming designed to optimize land productivity to feed areas with high population density and limited arable land.[56]

(Today, Chinese farmers apply mainly chemical fertilizers using digital and automated technologies. In 2007, Guangdong farmers utilized

---

[52] Kinver, Mark, "Growing pains of China's agricultural water needs" dated 24 June 2014 in BBC News [downloaded on 24 June 2014], available at https://www.bbc.com/news/science-environment-27978124.

[53] Huang, Jikun, Jun Yang and Scott Rozelle, *Op.cit.*

[54] Cook, Seth, "Sustainable agriculture in China: then and now" dated 26 March 2015 in International Institute for Environment and Development (IIED) [downloaded on 26 March 2015], available at https://www.iied.org/sustainable-agriculture-china-then-now.

[55] Francesca, Bray *Op.cit.*

[56] Cook, Seth *Op.cit.*

770 kg of fertilizer per hectare, two times the Japanese amount, five times more than that of Thailand, and six times the US figure, but a US$200 million initiative with World Bank (WB) assistance enabled Guangdong farmers to gain benefits from new sustainable agricultural technologies.[57])

Ji Yanyu improved the application of large amounts of several types of fertilizer on his Guangdong rice fields, which are less than half a hectare, but surplus chemicals flowed into groundwater or became residue on the crops.[58] Now, Ji utilizes only one type of fertilizer and a small percentage of pesticides compared to the past, resulting in an increase in rice yields from 450 kg in 2014 to 350 kg in 2013: "The new pesticides are much better than the old ones. Same thing with fertilizer. I used three to four types of fertilizers before. Now I use only one. This formula is good, very effective. The new pesticides are much better than the old ones. Same thing with fertilizer. I used three to four types of fertilizers before. Now I use only one. This formula is good, very effective."[59]

Concluding Remarks

In this chapter, we have discussed a brief historical survey of China's technological development and application in the agricultural sector. In each phase of the world historical development of agriculture, China either kept up with global trends or was leading them until the Industrial Revolution of the West overtook China in technological development. A gap then began to emerge in the modern era, as mechanization became highly developed in the West, while the Chinese agricultural sector lagged behind. Some observers argue that, by the era of the People's Republic of China (PRC) in 1949, almost all arable land was cultivated, with irrigation/drainage systems constructed a few hundred years before that, though intensive farming had relatively high outputs.[60]

---

[57] World Bank Group, "Technology Drives Sustainable Agricultural Development in China" dated 23 October 2015 in The World Bank [downloaded on 23 October 2023], available at https://www.worldbank.org/en/news/feature/2015/10/23/technology-drives-sustainable-agricultural-development-in-china.

[58] *Ibid.*

[59] *Ibid.*

[60] Talal Abu-Ghazaleh Global (TAG, Global), *Op.cit.*

The following chapter will examine how the PRC first contrasted with a thoroughly modernized West in the 20th century but began to accelerate its development after the 1978 economic reforms when it had access to the technologies found in the international community and the West. Economically, it developed speedily and then caught up with the West and even surpassed the West in the application of cutting-edge technologies. Chapter 4 is essentially a story of how the PRC overcame technological laggardness and propelled itself to the advanced frontlines of the contemporary agro-technological revolution.

Chapter 4

The Political Economy of Chinese Agricultural Sector

Tai Wei Lim

Abstract

This chapter on China briefly surveys the history of technological use in Chinese agriculture and food sectors. It adopts area studies and political economic angles to examine this topical matter with a focus on policy analysis related to the utilization of Industry 4.0 technologies.

Introduction

In the early days of the People's Republic of China (PRC), a country that just recently emerged from World War II (1937–1945), followed by the Chinese civil war (1945–1949), the PRC needed to feed a weary but hopeful population as one of the main priorities of its nation-building efforts. But few prime virgin lands could support population growth and economic development, with decline exacerbated due to the Great Leap Forward (1958–1960), a situation only reversed with the 1980s agricultural reforms that led to higher yields and bigger agricultural production from existing cultivated land.[1]

---

[1] Talal Abu-Ghazaleh Global (TAG, Global), "History of Agriculture in China" undated in TAG Confucius Institute (TAG-Confucius) website [downloaded on 1 January 2023], available at https://tagconfucius.com/article/view/History-of-Agriculture-in-China%C2%A0.

In the first phase of the reform period (1979–1984), the household responsibility system (HRS) significantly improved productivity and, from the mid-1980s, rural township and village enterprises' (TVEs) development created better market conditions through domestic market reform, fiscal and financial augmentation, exchange rate devaluation, trade liberalization, expansion of special economic zones (SEZs) to lure in foreign direct investments (FDIs).[2] This reversed the previous system adopted by the People's Republic of China (PRC). In the pre-reform era, when farming families were re-organized into cooperatives, collectives, and finally people's communes, incrementally.[3]

Chinese farmers began to be infected by market capitalism. From the early 1980s, the Chinese mass media vigorously promoted rural development with the state slogan "Getting rich is glorious," and the mainstream mass media comprehensively covered stories on villages, counties, corporations, and individuals who became wealthy from farming.[4]

According to the Food and Agriculture Organization (FAO) in 2002, the number of Chinese individuals who experienced malnutrition in the People's Republic of China (PRC) dropped from 193 million in 1990/1992 to 116 million in 1997/1999 (from 16% to 9% in the overall population).[5] The household responsibility system (HRS) transformed the collective agricultural production system to individual farms by farming out land-use rights to individual rural families and the creation of off-farm employment. By 2003, approximately 50% of China's rural labor force derived at least a component of its income from off-farm jobs, and non-agricultural income exceeded agricultural income in 2000.[6]

---

[2] Huang, Jikun, Jun Yang and Scott Rozelle, "China's rapid economic growth and its implications for agriculture and food security in China and the rest of the world" undated in the Food and Agriculture Organization (FAO) [downloaded on 1 January 2023], available at https://www.fao.org/3/ag088e/AG088E03.htm.

[3] Talal Abu-Ghazaleh Global (TAG, Global), *Op.cit.*

[4] Adusumalli, Harshini Priya, Nur Mohammad Ali Chisty, Rahman, Mahofuzur, Hossain, Shakawat Hossain and Pasupuleti, Mahesh Babu, "Bio-innovation and technological digitization of mushroom cultivation and marketing for rural development" dated 2022 in Academy of Marketing Studies Journal, 26(S3) [downloaded on 31 Dec 2022], available at https://www.abacademies.org/articles/bioinnovation-and-technological-digitization-of-mushroom-cultivation-and-marketing-for-rural-development-14654.html, pp. 1–7. {unpaginated online version was reference for this writing}.

[5] Huang, Jikun, Jun Yang and Scott Rozelle, *Op.cit.*

[6] *Ibid.*

A 1% boost in agricultural share in the GDP will result in almost 1% reduction of poverty incidence because off-farm employment is very significant for Chinese poverty reduction.

When the Chinese economic reforms first started in 1978, the household responsibility system (HRS) also began in the late 1970s to build a public agricultural extension system (PAES) and extension service centers in all rural counties and municipalities by the mid-1980s and, in 1985, to bring down public expenditures for the PAES, the central government motivated PAES to get their own income streams through retailing farm products.[7] PAES provides new technologies, information, human resource management, and capital to rural areas to set up new businesses, construct agricultural technology parks and industrialization bases, training and extension of agricultural machinery technology (encompassing trials, demonstration, and on-farm technical advice).[8] China in the late 1980s was positioned to tackle increasing demands for agricultural production with by hybridizing time-tested agricultural methods and modern agro-technology.[9]

Early results of China's agricultural industrialization were encouraging. The 1990s witnessed significant improvements in rural poverty alleviation. Based on the World Bank's (WB) US$1/day (in PPP terms) poverty line, rural poverty incidence declined from more than 30% in the early 1990s to approximately 8% in 2004.[10] Measured by China's official poverty line, more than 230 million Chinese rural residents have gotten out of poverty, and absolute poverty declined from 260 million in 1978 to less than 30 million in 2003, and rural poverty fell from 32.9% in 1978 to below 3% in 2003.[11]

At the same time, the 1978 market reforms accelerated urbanization and industrialization in China, starting from the Special Economic Zones

---

[7] Organization for Economic Cooperation and Development (OECD) and OECDiLibrary, "Chapter 6. The agricultural innovation system in China" dated 4 Oct 2018 in OECD Food and Agricultural Reviews, Innovation, Agricultural Productivity and Sustainability in China [downloaded on 1 November 2023], available at https://www.oecd-ilibrary.org/sites/9789264085299-9-en/index.html?itemId=/content/component/9789264085299-9-en [unpaginated online version].

[8] *Ibid.*

[9] Talal Abu-Ghazaleh Global (TAG, Global), *Op.cit.*

[10] Huang, Jikun, Jun Yang and Scott Rozelle, *Op.cit.*

[11] *Ibid.*

(SEZs) and then percolating throughout the country. China's industrialization was so remarkable that it was nicknamed the "factory of the world" when its southern coastal provinces started producing large numbers of consumer goods for the world. The "factory of the world" remains so today, except that its products are now going up the value-added chain and have to compete harder with rising "factories of the world" in India and Southeast Asia, where costs are lower than those of China. Agricultural makeup in China's overall economic output dropped from approximately 40% in 1970 and 30% in 1980 to 15% in 2004, while the services sector rose and increasing income in the early years of economic reform increased domestic demand, and the high savings rate resulted in physical capital investments in non-agricultural sectors in rural/urban areas.[12]

Contemporary China's farming industry is sizable. Alongside rapid urbanization, agriculture declined from in excess of 40% of the GDP in 1970 to 15% in 2004, while crop output upgraded to higher-value crops.[13] China has become the largest cultivator and consumer of a wide spectrum of major crops today, and it is importing increasing amounts of grains, soybeans, and other agricultural items, and a large-scale exporter of both fresh produce and processed foods.[14] China has come a long way. It is also fully integrated into the international community and the International Organizations (IOs).

Alongside its alignment with the rules and regulations of International Organizations (IOs), China has partnered with IOs like the FAO, UNDP, EU, ASEAN, CGIAR, CABI, PPIC, IAEA, IFS and WWF while engaged in regional agricultural with ASEAN, China, Japan and Korea (under the rubric of ASEAN+3), agricultural co-operation in Shanghai Co-operation Organization (SCO), China-FAO South-South Co-operation within the framework of the "Special Program for Food Security", and agricultural co-operation forums between China, central/eastern European countries.[15]

---

[12]*Ibid.*

[13]*Ibid.*

[14]Cook, Seth, "Sustainable agriculture in China: then and now" dated 26 March 2015 in International Institute for Environment and Development (IIED) [downloaded on 26 March 2015], available at https://www.iied.org/sustainable-agriculture-china-then-now.

[15]Organization for Economic Cooperation and Development (OECD) and OECDiLibrary, *Op.cit.*

It has approximately 2 million registered specialized farmers' cooperatives (as of the end of November 2018), which is 76 times larger than the number 10 years ago, and the cooperatives have more than 100 million rural households engaged in them (46.8% the national overall number). The specialized cooperatives' cultivators participate in the same category of agricultural production and therefore, they are able to combine resources and improve productivity, as Liu Xiaoran (a China National Association of Grain Sector specialist) explained: "The development of farmers' cooperatives in China will accelerate its land circulation, heralding greater demands for agricultural machinery and fierce competition among equipment manufacturers home and abroad."[16]

Even before Industry 4.0, China had put in place mechanical machines from Industry 3.0. As the year 2016 ended, China had 1.14 million combine harvesters, 14.31 million items of irrigation and drainage equipment (an increase of 105.3% and 6.1% respectively from 2006, data from China's 3rd national agricultural survey).[17] Some of these technologies came from the West. China's large consumer market lured non-Chinese producers of agri-tech to compete for its consumption, like America's John Deere and AGCO, and Italy's SDF Group, all of which are participants in Chinese machinery exhibitions and agricultural fairs.[18]

The bureaucracy is leading the charge in adopting agricultural technologies in China. Li Weiguo from the Ministry of Agriculture and Rural Affairs revealed China would push forward with the scientific and technological innovation of agricultural technologies with efficient machines and constant augmentation of mechanization.[19]

Agricultural R&D in China

Tech upgrades accounted for the biggest share of crop production growth even during the initial reform era, benefiting from the 1960s' rapid growth of China's research institutions, from nearly zero base in the 1950s, with contemporary capabilities of generating consistent output

---

[16] Xinhua, "Technology reshaping agriculture in China" dated 21 March 2018 edited by Huaxia in Xinhuanet [downloaded on 21 March 2018], available at http://www.xinhuanet.com/english/2018-03/21/c_137054369.htm.

[17] *Ibid.*

[18] *Ibid.*

[19] *Ibid.*

of semi-dwarf crop varieties before the global emergence of Green Revolution technology.[20] China introduced its agricultural innovation system in stages: (1) setting up the household responsibility system in place of the collective farming system and instituting individual land contract rights; (2) eliminated agricultural taxes and fees, put in place a direct payment system and augmented rural public services; (3) reform of the rural land system, bring online new agribusinesses and construction of new agricultural operation system like Famer Professional Co-operatives and Specialized Custom Plowers, Planters and Harvesters.[21]

China is adopting an all-of-society and all-of-government approach in managing R&D. The agricultural innovation system (AIS) in China gathers the state institutions, public research institutions, industry, university community, colleges, private sector firms, intermediate institutions, family farms, farmer co-operatives, agricultural enterprises, and other agribusinesses to form an ecosystem.[22] The 21st century brought China's status in the global agricultural revolution to new heights, with China's share in international agri-food publications going up from 1% to 9% between 1996 and 2012.[23]

Public investments in agricultural biotechnology was on the uptrend since the mid-1980s and, by 2003, the agricultural biotechnology research budget hit RMB1.6 billion (approximately US$200 million according to the official exchange rate or more than US$800 million in PPP terms) and a national research reform in the mid-1980s accelerated after the late 1990s to increase research productivity by moving from funding from institutional support to competitive grants, supporting research useful for economic development and encouraging applied research institutes to sell the technology they produce.[24]

In the 21st century, as China ramped up its public investments in agricultural R&D, agricultural R&D systems' output has progressed dramatically with its share in global agri-food publication rose from 1% in 1996 to 9% in 2012, alongside a rapid rise in agricultural patents

---

[20] Huang, Jikun, Jun Yang and Scott Rozelle, *Op.cit.*

[21] Organization for Economic Cooperation and Development (OECD) and OECDiLibrary, *Op.cit.*

[22] *Ibid.*

[23] *Ibid.*

[24] Huang, Jikun, Jun Yang and Scott Rozelle, *Op.cit.*

application and new plant variety rights, recording a yearly average growth rate of 26% between 2006 and 2011.[25]

The major source of public investments in agricultural R&D are from the state which has different priorities at various administrative hierarchical levels (national, provincial and municipal) or are sourced from state-related institutions like the National Natural Science Foundation, the Major National Science and Technology Projects, the National Key Research and Development Plans, the Guiding Projects for Technology Innovation, or the Innovation Base Projects and Talent Projects.[26]

The National Development and Reform Commission (NDRC) and the Ministry of Finance (MOF) dispense public funding for innovation and the technological upgrading of various economic sectors, while the National Natural Science Foundation of China coordinates resources for scientific research, for the state as it invests in agricultural innovation (including those with long-term returns and areas where there is a big disparity between social and private returns on investment).[27] In recent history, public investments in agricultural R&D sped up in the 2000s in funding expenditure and R&D personnel, expanding by a factor of almost 4 times between 2000 and 2013 in real terms, and, during that time, expenditure for public agricultural R&D in China surpassed that of the United States, making China the biggest investor in public agricultural research.[28]

Experimental test beds also exist in China. Agricultural science and technology experimental stations implement research outputs in various agricultural ecological areas as demonstrative experiments, while agricultural R&D divisions of private enterprises carry out market-driven independent research, production, and studies to enhance the main agricultural and forestry processed products.[29] There is a division of labor between public agricultural R&D institutions at the national, provincial, and municipality-level experimental stations and enterprises' technology R&D centers.[30] The national agricultural R&D institutions carry out major

---

[25] Organization for Economic Cooperation and Development (OECD) and OECDiLibrary, *Op.cit.*

[26] *Ibid.*

[27] *Ibid.*

[28] *Ibid.*

[29] *Ibid.*

[30] *Ibid.*

technologies research, high- and new technologies with strategic importance, basic and applied agricultural research, and general basic scientific/technological research work, while regional agricultural research centers work on major technologies with broad application for regional industrial development, applied basic and frontier research where China is proficient and has expertise, integrating and transferring technologies.[31]

One example of local research work carried out with multiple stakeholders is as follows. In the year 1999 in rural China, several groups of female peasants in a few rural villages, two plant-breeding groups and the Center for Chinese Agricultural Policy (CCAP, China's ranking public agricultural policy research organization) initiated a research project in Guangxi Province south-western China (well-known as a climatic/disaster-vulnerable mountainous region) where the villagers-peasants cultivate maize in small pockets of soil on steep slopes and between rocks in flat fields.[32] The topography of this region resulted in limited irrigable water, and rains tended to result in floods and wash away crops in addition to the fact that there are no roads and access points to markets is restrictive.[33]

Those are the conditions that some local communities have had to work with. In such areas, the traditional staple crop grown for consumption is maize, where there is a diversity of maize landraces, but in comparatively more well-off valley and flatland communities where residents have higher education and are more market-conscious, maize has become pig feed as pig farming takes over as the major source of village income.[34] CCAP senior researcher Yiching Song (doctorate in rural development from Wageningen University, Netherlands) came up with the ambitious initiative's agenda of crop rotation, organic fertilizers usage, and participatory research to synergize between farmers and breeders while coming up with innovations in utilizing traditional and modern knowledge/practices.[35]

---

[31] *Ibid.*

[32] Vernooy, Ronnie, "China Tries Alternative to Industrial Agriculture" dated 20 July 2012 in United Nations University (UNU) [downloaded on 20 July 2012], available at https://ourworld.unu.edu/en/china-tries-alternative-to-industrial-agriculture.

[33] *Ibid.*

[34] *Ibid.*

[35] *Ibid.*

The research team (including farmers) tapped into local community female maize-breeders' long years of experience and qualified professional plant breeders to set up participatory plant breeding collaborations between the seed production and farming systems to enable mainly female farmers' incomes to increase crop yields by combining scientific (crossing techniques and variety selection processes) and traditional knowledge.[36] Breeders applied more advanced methodologies like those developed in the Guangxi Maize Research Institute (GMRI) Nanning fields to produce 80 higher-yielding species that can withstand biotic and abiotic stresses (e.g., pests, diseases, and drought) and, after a decade of experimentation, 4 peasant-endorsed species were shortlisted and grown first in the research villages and then in the proximate Yunnan and Guizhou provinces.[37]

One of the stakeholders in this project, Yiching Song, opined: "Our work in Guangxi tells us that we need to work together with scientists from many disciplines, and with multiple stakeholders from multiple levels. … It also shows that concrete action can bring about more profound innovation through better understanding of and dealing with the complexity of rural realities, step by step. It takes a long time, but it is possible."[38] The research team motivated farmers and breeders to sign a formal contract regarding sharing the breeding materials, seed production methods, and marketing local community crops with the official knowledge of the Ministries of Agriculture and Environmental Protection against the backdrop of global agreements like the Convention on Biological Diversity and the International Treaty on Plant Genetic Resources for Food and Agriculture.[39]

Legal Aspects of Innovation Policies

The Chinese state has also tweaked the legal aspects of innovation policies in China to encourage agricultural R&D. To encourage R&D in agricultural machinery technology, the state promulgated the Law on Promotion of Agricultural Mechanization in 2004 with stipulations

---

[36] *Ibid.*

[37] *Ibid.*

[38] *Ibid.*

[39] *Ibid.*

supporting capital, projects, R&D policies, extension/social services of agricultural machinery.[40] While the laws, directives, and regulations are there, in China, legal enforcement can be effected via the judicial or administrative system, but both judicial and administrative decisions are challenging to implement due to the absence of legal infrastructure, mechanisms, and human resources, often encountering limitations and restrictions put in place by local governments.[41]

Other than R&D laws and regulations, IPR laws have also been augmented over the reform years and beyond. The contemporary Intellectual Property Rights (IPR) protection system in China was systematized after the late 1970s alongside Deng Xiaoping's economic reforms, and, from the 1980s, China promulgated IPR protection laws/regulations with the Patent Law first enacted in 1985 and then amended 3 further rounds thereafter.[42]

China also aligned its domestic laws with those of global institutions. China made attempts to comply with the regulations of international organizations (IOs). It signed the Madrid Agreement Concerning the International Registration of Marks in 1989, the Universal Copyright Convention in 1992, the Patent Co-operation Treaty (PCT) in 1994, amended its IPR laws before ascension into the WTO in 2001 to comply with the WTO Agreement on Trade-related Aspects of Intellectual Property Rights, and joined the International Convention for the Protection of New Varieties of Plants and World Intellectual Property Organization Copyright Treaty in 1999 and 2007, respectively.[43]

Irrigation Tech in Contemporary China

In the modern and contemporary eras, China has had to work with geographical challenges. Northern China has spacious arable land, but groundwater resources are limited, and dried river beds are increasing in number in China.[44] While groundwater extraction in northern Chinese

---

[40] Organization for Economic Cooperation and Development (OECD) and OECDiLibrary, *Op.cit.*

[41] *Ibid.*

[42] *Ibid.*

[43] *Ibid.*

[44] Kinver, Mark, "Growing pains of China's agricultural water needs" dated 24 June 2014 in BBC News [downloaded on 24 June 2014], available at https://www.bbc.com/news/science-environment-27978124.

provinces was unsustainable due to the speed of consumption racing faster than replenishment, less than 30% of the known groundwater resources in southern China were utilized despite its comparatively plentiful supply of surface water.[45]

Chinese officials labeled China's water shortage a "grave situation" and urged tighter water management, as they were afraid that it could undo attempts to attain sustainable development.[46] In 2012, China's vice minister of water resources, Hu Siyi highlighted the following: "Because of the grave situation, we must put in place the strictest water resources management system…[about two-thirds of Chinese cities were "water-needy", nearly 300 million rural residents lacked access to safe drinking water, and 40% of rivers were seriously polluted]".[47]

In addition, in 2014, a scientific research team from the US, Japan, and China evaluated the amount of water utilized by various provinces to cultivate crops and livestock, rainwater volume, and irrigation systems, and they concluded China's scarce water supply was used non-optimally, as crops grown in water-scarce provinces were diverted to humid, rainfall-heavy regions.[48]

The international research team made up of US, Japanese, and Chinese personnel highlighted in 2014 that: "China faces most of the major challenges to sustainable agriculture. Fast socioeconomic development, rapid urbanisation and climate change, along with very limited water resources and arable land per capita. [Because arable land is available mainly in the water-scarce north, irrigation has become widespread, covering 45% of the country's agricultural land and accounting for 65% of national water withdrawal]. China's domestic food trade is efficient in terms of rainwater but inefficient regarding irrigation, meaning that dry, irrigation-intensive provinces tend to export to wetter, less irrigation-intensive ones. We (also) identify specific provinces (for example, Inner Mongolia) and products (for example, corn) that show high potential for irrigation productivity improvements. They constitute an essential input for designing policies and provide a framework for analysing how these policies might change China's… irrigation use in the near future."[49]

---

[45] *Ibid.*

[46] *Ibid.*

[47] *Ibid.*

[48] *Ibid.*

[49] *Ibid.*

Scalability

Chinese rice fields were very small-scale (often becoming smaller irrigation units), utilizing mainly small, light, and inexpensive farm implements with water buffalos as beasts of burden, a chain-pump or small portable wood-and-bamboo pumps, and they did not grow at the same rate as Western farms and machinery, which were rapidly expanding in size in the modern and contemporary eras.[50] There were a number of reasons why Chinese farms did not want to expand in size. Small-scale Chinese farmers did not have the budgets for advanced machines, and large combine harvesters were not appropriate for small mixed crop fields, as they did not have the economies of scale of large US-style monoculture farms that have taken root in the modern world.[51] The fragmentation of Chinese agricultural production into a large number of small companies producing multiple homogeneous products means that there is only a small number of farms willing to invest in private R&D activities, as they are not keen to invest without being confident of recovering the investment.[52]

The economies of scale are increasingly important due to the growing appetite for all kinds of foods in China. In 2014, an international team of American, Japanese and Chinese researchers' findings were published in the Proceedings of the National Academy of Sciences focusing on four major crops (soya, wheat, rice and corn or maize) and three livestock meats (ruminant, pork and poultry) as they make up 93% of China's domestic food supply in 2005 (according to the United Nations/UN).[53]

While scaling up farms may result in a greater capacity to feed more people with larger harvests, there were, however, some advantages of keeping farms small too. Smaller farmers meant individual peasants could exercise greater accurate control and management of the land with basic technologies to improve productivity without needing massive expansion

---

[50] Francesca, Bray, "Rice, Technology, and History: The Case of China" dated Winter 2004 in Education About Asia, Association for Asian Studies (AAS) Education About Asia: Online Archives, Volume 09:3 (Winter 2004): Special Section on Teaching About Asia Through Rice) [downloaded on 1 January 2023], available at https://www.asianstudies. org/publications/eaa/archives/rice-technology-and-history-the-case-of-china/.

[51] *Ibid.*

[52] Organization for Economic Cooperation and Development (OECD) and OECDiLibrary, *Op.cit.*

[53] Kinver, Mark, *Op.cit.*

of labor.[54] From 2008 to 2018, China set up large-scale agricultural machinery for the northern and north-eastern plains.[55] Jilin is an example of a location in China where the technological take-up rate has been high. The extent of Northeast China, Jilin province's adoption of mechanization for cultivating its major crops is above 80% of its farms, and the province owns approximately 600,000 large and medium-sized tractors.[56]

Policy Regarding the Use of Industry 4.0 Technologies

Contemporary China has 128 million hectares of arable land, with approximately 50% of it slated as high-quality farmland for grain and other prioritized crops to feed 20% of the global population.[57] Boosts in food production have bumped up Chinese food consumption per capita, with daily per capita food availability per day increased from 1,717 kcal in the early 1960s to more than 3,000 kcal by the late 1990s.[58] Technology is seen as the great solution for feeding China's population.

The tech factor features importantly in China's five-year economic plans (FYPs). The 2016 No. 1 Document of the State Council stated innovation as one of the topmost agricultural policy priorities to bring about and promote healthy, sustainable agricultural growth that is innovative, coordinated, and green, while the 2017 No. 1 Document prioritized basic research in agricultural R&D and augmenting innovation capacity through building national agricultural high-tech development zones, while pushing for public/private partnerships.[59]

The 2017 No. 1 Document also augmented agricultural R&D, emphasizing public support for inclusive, diverse actors in agricultural science

---

[54] Francesca, Bray, *Op.cit.*

[55] Organization for Economic Cooperation and Development (OECD) and OECDiLibrary, *Op.cit.*

[56] Xinhua, *Op.cit.*

[57] Khan, Omar, "China Agriculture Innovation: Advanced technology and new methods bring stability to agricultural industries, help improve food security" dated 22 Feb 2023 in the CGTN [downloaded on 22 Feb 2023], available at https://news.cgtn.com/news/2023-02-22/VHJhbnNjcmlwdDcwNjky/index.html.

[58] Huang, Jikun, Jun Yang and Scott Rozelle, *Op.cit.*

[59] Organization for Economic Cooperation and Development (OECD) and OECDiLibrary, *Op.cit.*

and technology by strengthening incentives for agricultural innovation, having researchers in public research institutions work in the private sector, instituting results-based compensation, putting in place a differentiated technology evaluation system, and improving agricultural intellectual property rights.[60]

Scientific workers in China appear to agree with their government's FYP focus on science and technology. Yin Yulong (Principal Investigator from the Institute of Subtropical Agriculture, Chinese Academy of Sciences, or CAS) opined: "Modern science and technology, such as intelligent breeding are crucial. We can save on labor and improve efficiency. If we use robots to raise livestock, we reduce contact with humans, therefore reducing the occurrence of diseases. Biotechnology is also important. Using fermented microorganisms as feed additives can improve the quality of agricultural products."[61]

Climatic and Temperature Control

Chinese agricultural tech is used to track weather in paddy fields (a rice tech demo park is set up in Fujin City in the northeast Chinese province of Heilongjiang), managing watering and fertilizing schedules, cutting agricultural attrition by diagnosing crop diseases and pests.[62] Farmers there noted that cherry tomato crops are prone to fungal diseases, so they use technological means to evaluate the extent of fungal spores dispersed in the air and channel real-time data to the cloud in order to alert technicians of the need to track and resolve unusual spore-related situations.[63] This is an early warning system that alerts farmers of the need to nip fungal outbreaks in the bud before they become full-blown infections.

There were other ways that technologies were used to shape the climate conditions for the crops. In terms of temperature-controlled facilities, Wen Yesheng (leader of Zhongyi agricultural machinery

---

[60] *Ibid.*

[61] Khan, Omar, *Op.cit.*

[62] Xinhua, "Smart farming advances China's modern agricultural development" dated 17 Feb 2023 edited by Huaxia in Xinhuanet [downloaded on 17 Feb 2013], available at https://english.news.cn/20230217/975b5bb7f7a345cdb83530dae9a33d4d/c.html.

[63] *Ibid.*

cooperative)'s high-tech greenhouse no longer required him to lift up the curtains to decrease the temperature combust coal for heating.[64] The accent of such technologies was to increase yield and reduce harvest losses. In addition to mechanized curtains, the facility also has a mechanized exhaust system and solar energy equipment, all of which can assist Chinese cultivators in cutting down on grain losses in the harvesting stage.[65]

Robotic Use

Labor supply for agricultural industries is declining as large numbers of rural peasants become economic migrants to the urban cities.[66] The utilization of robots can be one solution to this challenge, especially since Industry 4.0's development of Artificial Intelligence (AI) facilitates the development of semi-autonomous (and eventually autonomous) smart robots. Smart robots also harvest mature fruit and vegetables using imaging technology to gauge the distance between the fruits for picking, and their robotic arm, and then execute the assignment in under a second for efficient/fast-paced work. Since the end of 2021, a Chinese robotic industry development plan emphasized functions like weeding, fruit picking, poultry feeding, and sludge cleaning.[67]

In the large-scale grain-cultivating Nong'an country in northeast China's Jilin Province, drones were utilized to release white mist of chemicals on the crops, as Wen Yesheng (head of Zhongyi agricultural machinery cooperative) explained: "Drone spraying is five times more efficient than tractor sprayers, let alone manual spraying. That's the power of modern agriculture."[68] Drone users for pesticide spraying can also be found in the wheat-growing facilities of Nanyangjialou Village, Longyao County, in Xingtai City, Hebei, northern Chinese Province.[69]

---

[64] Xinhua, *Op.cit.*

[65] *Ibid.*

[66] Cook, Seth, *Op.cit.*

[67] Xinhua, *Op.cit.*

[68] Xinhua, *Op.cit.*

[69] *Ibid.*

Core Leader's Message for Agricultural Technology Use

The leading group of science and technology education of the State Council manages the innovation system and makes major decisions for the most important innovation plans, including coordinating national ministries and agencies while the Ministry of Science and Technology (MOST) and the State Intellectual Property Office develop R & D innovation policy (including agriculture blueprints), then the Ministry of Agriculture (MOA), Ministry of Water Resources (MWR) and the State Forestry Administration translate these blueprints into practice through formulating the agricultural R&D policies, alongside consultations with public R&D institutions.[70] But the political decisions come from the very top echelons of the Politburo and the Politburo Standing Committee (PSC) of the Chinese Communist Party (CCP), and in the era of President Xi Jinping, the core leader has personal oversight of the process.

In December 2022, Chinese President, General Secretary of the Communist Party of China (CPC) Central Committee, and chairman of the Central Military Commission (CMC), Xi Jinping, spoke at the yearly central rural work conference in Beijing, the capital of China, from December 23–24. Xi noted in the following transcript on state-owned press:

> Advancing rural revitalization across the board and moving faster to build up China's strength in agriculture is part of the CPC Central Committee's strategic plans of building China into a great modern socialist country in all respects. Xi called for full steam in pressing ahead with the work related to agriculture, rural areas and farmers with rural revitalization as the focus, and vigorously promoting agricultural and rural modernization.
>
> In building up China's strength in agriculture, efforts should be made to reflect Chinese characteristics, and take into consideration China's national conditions, resource endowment featuring limited land for a huge population, the historical background of farming civilization, and the times' requirement of harmony between humanity and nature.
>
> China should take its own path instead of simply following the foreign models of modern agriculture development . . . strengthening

---

[70]Organization for Economic Cooperation and Development (OECD) and OECDiLibrary, *Op.cit.*

top-level design and developing plans to move faster to build up China's strength in agriculture, taking steady and incremental steps, and solving practical issues that are most urgent for agricultural and rural development, and of the most pressing concern to farmers.

It is always a priority to ensure a stable and secure supply of grain and key farm produce . . . efforts to keep the country's farmland area above the red line of 1.8 billion mu (about 120 million hectares) and effectively revitalize the seed industry . . . efforts should both be made to increase grain output and reduce losses, as well as to establish a diversified food supply system and expand food sources in multiple ways.

Advancing rural revitalization across the board is a crucial task for building up China's strength in agriculture in the new era . . . demanded shifting manpower, material resources and fiscal support to rural revitalization, saying consolidating and expanding the achievements in poverty alleviation is the bottom line task of rural revitalization . . . the need to rely on both science and technology and reform to facilitate the work of strengthening China's agriculture . . . efforts to establish a multi-level agricultural science and technology innovation system with sound division of labor and cooperation, and moderate competition . . . deepening rural reform so that more fruits of reform could be shared by farmers.

It's necessary to remove institutional barriers that prevent the equal exchanges and two-way flows of production factors between urban and rural areas, and facilitate the flow of development factors as well as various services into rural areas . . . modernization of rural areas is the intrinsic requirement and necessary condition for building up China's strength in agriculture, which entails the building of a beautiful and harmonious countryside that is desirable to live and work in . . . faster development of public service facilities in fields like epidemic prevention, elderly care, education and medical care in rural areas. Efforts are needed for improvement in rural infrastructure, public services and living environment to ensure that residents in the countryside can lead modern lives. The Party's leadership must be upheld unswervingly in the work related to agriculture, rural areas and farmers.[71]

---

[71] Xinhua, "Xi Focus: Xi stresses building up China's strength in agriculture at key rural work conference" dated 24 Dec 2022 in Xinhuanet edited by Huaxia [downloaded on 24 Dec 2022], available at https://english.news.cn/20221224/e0cb406446c94f60a301e 86a0a1b88aa/c.html.

Former Premier Li Keqiang presided over the conference (with the participation of Li Qiang, Wang Huning, Han Zheng, Cai Qi and Ding Xuexiang) and Li asked the conference to seriously read Xi's significant policy talk, actualize decisions and plans made by the central government while former politburo member Hu Chunhua said Xi's talk is a scientific guide/blueprint of action for augmenting China's strength in agriculture, hastening agricultural/rural modernization, and promoting rural revitalization nationally.[72]

With increasing automation and digitalization, Chinese farmers are moving away from farming based on precedents and experience to place greater reliance on digital tech, and this is also encouraged officially in the 2023 "No.1 central document" that augments official support for agricultural science, technology and equipment by encouraging rural digital development, exploring various scenarios for digital technology implementation, quickening the pace of big data development, and pushing forward with smart agriculture.[73]

Participation of the Private Sector

Private enterprises within the same industry or industry chain organize associations or alliances to implement joint R&D programs, mutually study each other's best practices, tailor resources to individual enterprise and, as middlemen, they can facilitate the flow of knowledge, capital, information, and talents, and align supply and demand needs of agricultural innovation.[74]

Non-government funds from donors, private investment in agriculture R&D by enterprises, financial institutions and venture capital are sources of finance for basic and non-profit agricultural research and the commonality amongst them is their agricultural R&D activities have a high market return on investment, typically focusing on applied and development research or projects that protect their returns through the observance of intellectual property rights, e.g., areas like food processing, agricultural chemicals, farm machinery, hybrid seeds and genetically modified crops.[75]

---

[72] *Ibid.*

[73] Xinhua, *Op.cit.*

[74] Organization for Economic Cooperation and Development (OECD) and OECDiLibrary, *Op.cit.*

[75] *Ibid.*

The Chinese private sector is also involved in the quest to tap into Industry 4.0. Evolving from managing agricultural spaces to a crop breeding lab, Xiangyan Seed Industry advanced agriculture firm, is keen to push for science and new technology to revolutionize farming, especially for the crops of peppers, corn, eggplants, and other vegetables.[76]

On the other hand, some argue that excessive privatization of seed production, emphasizing hybrids and modern species, led to neglect of traditional and underutilized crops, but wide hybrid varieties had difficulties adapting to the conditions in remote mountainous areas, including coping with more inconsistent weather conditions (along with natural disasters like droughts, floods), in Guangxi and other south-western provinces.[77] The same individuals argued that the uniform application of privatized production of seeds appears to be vulnerable to diseases and pests, thus some cultivators/farmers in these southern areas were not able to accept hybrid species and remained reliant on domestic species.[78]

Local Area Adoption of Agricultural Technologies

From the highest economic growth rates reached in 1984 (15%) to the lower but still rapid growth into the late 1980s, economic reform widened from agricultural to non-agricultural sectors, local leaders gathered capital and labor to construct rural township and village enterprises (TVEs), especially after the mid-1980s, alongside general liberalization in many sectors.[79] Just as the local leaders semi-urbanized the rural areas, they were also an important leading stakeholder in implementing Industry 4.0 technologies and innovation in the rural regions as well.

Besides the national landscape in terms of technology adoption, local cities and counties are also tapping into Industry 4.0. Local authorities at the provincial levels, parallel to central state organizations and their Science and Technology Commissions (including the minority autonomous regions and municipalities), play a major role in policy making, implementation, and financing R&D and innovation at the local level.[80]

---

[76] Khan, Omar, *Op.cit.*

[77] Vernooy, Ronnie, *Op.cit.*

[78] *Ibid.*

[79] Huang, Jikun, Jun Yang and Scott Rozelle, *Op.cit.*

[80] Organization for Economic Cooperation and Development (OECD) and OECDiLibrary, *Op.cit.*

The central government also works with the local authorities to draw up, apply, and operate agriculture innovation policies, grow the institutions, fund/carry out R&D (and extension activities), coaching/training activities, investments in agricultural activities, and construct infrastructure to support agricultural innovation.[81]

Local Farming Industry Use of Mechanization. Local governments and non-government organizations established agricultural machinery-leasing facilities, harvesting machinery service firms, and agricultural mechanization associations to offer a wide range of services concerning the operation of agricultural machines and filling in hardware provision found lacking in government services.[82] This is to meet some local demands for leasing needs, as some locales prefer to acquire their own equipment rather than leasing it due to various reasons.

In 2018, Minyue Village Songyuan City cooperative began utilizing an advanced corn harvester that can lacerate the crop and sieve off the seeds from the cob while processing and spreading the corn stover over the field, as village party chief Zhang Zhifeng indicated: "The traditional harvester collected corn and transported them to a storage house for further separation work. Around 1 percent of seeds was lost during the process. The new machine has higher efficiency and can greatly reduce the losses of harvest".

Local Farming Industry Use of Chemical Technologies. With the use of technologies, fertilizers, and pesticides, rice fields in Changsha County experienced almost 39% year-on-year (y-o-y) yield increases in yields due to an annual 50 million yuan pumped into the Changsha County seed industry's R&D projects.[83] Tang Qiyuan, a scientist in the China Agriculture Research System of Hunan Agricultural University, indicated how Changsha City "is building four bases for grain production, with one focus on advanced technology. It also includes high-quality and high-yield output. The aim is to improve the level of food production while also gradually introducing modern agricultural technologies to farmers. Over time, it's transformed from small to large scale."[84]

Local Farming Industry Use of Big Data. Zhou Shijie, a tomato farmer in Zhengding County, northern Chinese province of Hebei, is

---

[81] *Ibid.*

[82] *Ibid.*

[83] Khan, Omar, *Op.cit.*

[84] *Ibid.*

utilizing a refreshing methodology of using digital data to determine the schedule of watering and fertilizing his crops by reading from a monitor indicating real-time data derived from sensors in the high-tech tomato greenhouse: "Air humidity 55 percent, soil temperature 27.9 degrees Celsius, and soil pH 6.1. By looking at the data, I can tell when my plants are thirsty for water or hungry for fertilizer."[85]

Funding Training Needs

To disseminate knowledge, the state sets up training institutes to impart modern knowledge to farmers. They include the following institutions. Farmers' cooperatives provide coaching on technical issues like rice cultivation technology, local community farming schools are set up for those who lost land and retraining for new jobs in intensive agricultural production operations, rural-area evening schools are established as part of the "one village one product" initiative to focus on a local specialty product; while other evening classes teach courses on cultivating tobacco and hybrid rice cultivation, off-season vegetables cultivation and rice flour processing.[86]

To professionalize farmers' expertise, the central government funded training programs for specialized farmers, family farm managers, agricultural co-operatives heads, agribusinesses/agricultural service providers, migrant workers, rural talents, tertiary-educated rural leaders, vocational school students, and also the National Top Ten farmers' best practices program.[87]

With its increasing tech prowess, China began to adopt an external gaze as it trained more than 4,000 African agriculture technicians and management personnel from 2004 to 2018.[88] From at least 2008 to 2018, China started 150 agricultural exchanges/cooperative projects, and trained more than 1,000 ASEAN-based technical and managerial staff in courses on agricultural technology under the rubric of ASEAN+2 and ASEAN+3.[89]

---

[85] Xinhua, *Op.cit.*

[86] Organization for Economic Cooperation and Development (OECD) and OECDiLibrary, *Op.cit.*

[87] *Ibid.*

[88] *Ibid.*

[89] *Ibid.*

Sustainable Development and Pollution Mitigation

The Chinese Ministry of Agriculture and Rural Affairs data indicated mechanization of crop cultivation, planting and harvesting functions increased from 57% in 2012 to more than 72% in 2021 and beyond actual farming duties, the government is also helping cultivators sell their output on the internet, eventually driving retail sales to more than 800 billion yuan by 2025 (as stated in the national agriculture and rural informatization development plan for the 14th Five-Year Plan FYP, 2021–2025).[90]

In the past few decades, the proliferation of conventional agriculture has resulted in destructive outcomes, and this has driven China to focus on the revival of sustainable agriculture in China in recent years.[91] Increasingly, the accent is now on sustainable farming, managing declining conditions of arable lands, and the mitigation of pollution. The use of fertilizers and pesticides is not exceptional in China, but its fertilizers and pesticides applications are now among the highest in the world, resulting in soil erosion, soil pollution, decline in agricultural biodiversity, water scarcity in many areas, and decreasing water tables in northern China.

China's inaugural national pollution source survey in 2010 indicated agriculture (not industry) was the main reason behind surface water pollution.[92] Cao Wendao (World Bank senior rural development specialist and co-leader for the Guangdong Agricultural Pollution Control Project or GAPCP) said: "It has become a serious problem. We wanted to do it in a targeted way. After intensive discussion with local officials and technicians, we came up with the idea of the IC card system. And it is very effective."[93]

Fertilizers and Pesticides

Guangdong experts indicated that approximately 30–40% of pesticides reached their goal of eliminating pests, but the rest of the overflow seeped into the environment or became residue on agricultural products, affecting

---

[90] World Bank Group, "Technology Drives Sustainable Agricultural Development in China" dated 23 October 2015 in The World Bank [downloaded on 23 October 2023], available at https://www.worldbank.org/en/news/feature/2015/10/23/technology-drives-sustainable-agricultural-development-in-china.

[91] Cook, Seth, *Op.cit.*

[92] World Bank Group, *Op.cit.*

[93] *Ibid.*

food safety.[94] In other words, they became sources of pollution. The Guangdong Agricultural Pollution Control Project (GAPCP) started in early 2014 to tackle pollution from both large-scale livestock and crop production as an inaugural cooperation between the World Bank and Guangdong to dispense subsidies and technical support to more than 100,000 households to implement environmentally-sound crop production.[95] The project introduced high-quality pesticide equipment like improved sprayers, integrated pest management methods, natural pest predators, pest lamps, insect glue boards, drones, and other tech.[96]

To obtain fertilizers tailored to crop growth needs and high efficiency, low toxicity, and low residue pesticides, farmers utilized an integrated circuit card (IC or smart card) that has information on their name, land size, amount of subsidized fertilizers/pesticides, and subsidy amounts.[97] The farmers can acquire goods only in competitively chosen agriculture supply stores with card reader facilities and pay subsidized prices directly to ensure the subsidies are utilized as intended.[98]

Jiang Ru, the World Bank senior environmental specialist and project co-leader, indicated: "We hope positive results achieved through the project can help the province, and the rest of China, identify sustainable pathways for its agricultural sector."[99] With the project implementation, fertilizer application declined by 24% for the spring rice and 12% for autumn rice in 2014 while applied pesticide for rice dipped by 27%, while spring rice output expand 6% and autumn rice gained 19%.[100]

By cutting down pesticides and fertilizers used, farmers no longer pollute local water systems or saturate their products with surplus chemicals, while a US$200-million project supported by the World Bank facilitates pig farmers' recycling of animal dung into household energy fuel without polluting local water supplies.[101] Besides crop cultivation techniques, there are also recycling and environmentally-friendly

---

[94] *Ibid.*

[95] *Ibid.*

[96] *Ibid.*

[97] *Ibid.*

[98] *Ibid.*

[99] *Ibid.*

[100] *Ibid.*

[101] *Ibid.*

technologies and management techniques in place in China to handle agricultural processes.

Even animal wastes are recycled as fertilizer compost in hog farms. Some newly-built Guangdong farms have two-levels with the hogs reared on the second storey and their dung and urine removed via a slatted floor to the first storey, and they are then composted as organic fertilizer (unlike the traditional single-floor farms where pig wastes are flushed into water bodies).[102] Zhang Zhiqiang's Guangdong-based pig farm processed the discharge of 4,000 pigs, creating pollution and a nasty smell that offended his neighbors, producing environmentally acceptable treated water and biogas from the dung, which then powers the farm's electricity needs.[103]

On a smaller scale, non-governmental organizations (NGOs) were also involved in advocating the reduction of chemical fertilizers and pesticides. Two organic associations in the villages of Chentang and Sancha, set up in 2005 and converted to organic farming from industrialized farming, began their project with 10 farmers who got technical guidance and financial support from an HK grassroots poverty alleviation NGO Partners for Community Development (PCD). In 2006, PCD personnel collaborated with the Center for Chinese Agricultural Policy (CCAP) teams and Guangxi Maize Research Institute (GMRI) to enhance local crop species by encouraging regular communications between members to bring about better farming skills and aid greater comprehension of the merits of organic farming, while monitoring each other to stay away from chemical fertilizers and pesticides.[104] As they increased their organic farming spaces, domestic fertilizer supply could not satiate the demand, so ingredients like bran and bone meal were acquired for association members to experiment with making their own fertilizer based on local conditions under strict quality control.[105]

The Rise of Organic Farming Technology

Due to large-scale privatization and commercialization, the formal official Chinese seed system was incrementally exposed to profit-driven, aggressive market competition, leading to concerns about biodiversity and rural

---

[102] *Ibid.*

[103] *Ibid.*

[104] Vernooy, Ronnie, *Op.cit.*

[105] *Ibid.*

livelihoods.[106] Moreover, there are changing lifestyle trends in the more affluent parts of China. With rising incomes, the emergence of a middle class, and greater awareness of a healthy lifestyle, Chinese consumers now yearn for organic foods.

The first-known conversion to organic farming go back to early 1990s and, by 2002, organic vegetables appeared in some major supermarkets in big cities driven by increasing local and foreign demands which resulted in more than 500,000 hectares roped out for certified organic farm product cultivation operated by more than 1,000 companies by 2005 (and this trend eventually expanded nationwide to cover cash crops like tea and bamboo).[107] The earliest Guangxi organic products include rice and kohlrabi grown without the utilization of industrial inputs, and they were popular with Nanning, Liuzhou, and Hong Kong consumers and restaurants, e.g., Nanning's increasingly popular organic-food restaurants acquired such products from two organic associations with prices only marginally above those of conventional crops.[108] These two trends were thus clear: organic products were becoming popular with consumers, and prices were getting lower for organic products.

Southern China case studies

In southern China (generally considered the wealthier regions because the 1978 economic reforms started there), Organic or green farming associations set up by local farmers and supported by the Center for Chinese Agricultural Policy (CCAP, China's ranking public agricultural policy research organization) research team in Hengxian County (southeast Guangxi) are involved in green certified organic agriculture building upon the foundation of historical farming systems decoupled from overreliance on industrial inputs.[109]

Another example is found in Jiangxi. Wanzai County, founded in 1999 in Jiangxi Province after the County's People Congress voted to turn the whole township into an organic farming hub and kept out all synthetic chemical-based farming compounds, has become a ranking success story

---

106 *Ibid.*

107 *Ibid.*

108 *Ibid.*

109 *Ibid.*

in big-scale organic farming in China by having local authorities coach farmers in utilizing contemporary technologies and marketing techniques for their products.[110] The Wanzai farms are so successful that they have developed significant export potential. By the end of 2014, Wanzai had a total of 5,400 hectares of rice fields with organic certified rice, ginger, soybean, strawberry, scallions, yam and other cash crops serving the Chinese consumers and for selling to overseas customers.[111] Organic agriculture in Wanzai had evolved into the County's main industry output and developmental engine, necessitating a strategy mobilizing and/or employing 17,000 households from 48 villages in 11 towns.[112]

Organic farming runs into the challenge of the central government not endorsing any agricultural techniques that have the potential of lowering production, even with positive environmental impact and better farmers' incomes, as food security is the topmost priority for the state.[113] In addition, complex and overlapping eco-labeling benchmarks for sustainable agricultural foods have made it difficult to certify these foods, which runs into the problem of large numbers of Chinese people distrusting these items because of food scandals and negative press.[114]

Non-Government Organizations/Non-Profit Organizations (NGOs and NPOs)

Despite the state-centered Chinese approach to agricultural development, there are some non-government, non-profit organizations (NGO, NPOs) engaged in China's agricultural industry. For example, Shared Harvest, a community-supported agriculture (CSA) enterprise established in 2011 by two doctoral urbanite researchers with no prior farming experience, had 460 members working in the Beijing suburbs.[115]

Unlike the peasant-driven farming trends in the past few thousand years, Shared Harvest is an unusual counter-culture in the agricultural movement and development in China; thus, it is not surprising that there

---

[110] Cook, Seth, *Op.cit.*

[111] *Ibid.*

[112] *Ibid.*

[113] *Ibid.*

[114] *Ibid.*

[115] *Ibid.*

is strong mass media attention on the trend.[116] Developed into China's leading community-supported agricultural CSA enterprise, it is a demonstrative case of urban intellectuals engaged with farming who have environmental awareness and an enthusiasm for sustainable food supply.[117]

There are challenges for these CSAs. Shared Harvest utilizes completely organic farming methodologies, but is not officially certified by the authorities, and, as with most small-scale Chinese farmers engaged in sustainable farming methods, the cost factor of the certification process is restrictive and a major challenge.[118]

Other non-profits, like cooperatives, also perform the function of facilitating the authorities' agenda of implementing new technologies at the local level. Farmer co-operatives act as middlemen to facilitate the adoption of new technologies, lower transaction costs, assisting small farms to transcend systematic limitations in utilizing technologies, integrating them in supply chains, and increasing their operations.[119]

Concluding Remarks

Overall, due to land redistribution exercises, rural China's food accessibility has improved over time due to fairer distribution of lands to the peasants, introduction of market forces, state natural disaster assistance, increased opportunities in non-farm employment, state investments in the sector, construction of rural infrastructure, and rural township and village enterprises (TVEs).[120]

To maintain a harmonious society, access to stable food supplies by the lower-income individuals is an important anchor of food security, so the state has a disaster relief program, a national food-for-work scheme for disaster relief/long-term investments, as agriculture is a main source of income for the poor.[121] In Chinese history, when food security is disturbed by unrest, society begins to unravel, as did the last Chinese imperial dynasty, after the Qing granary system was mostly damaged/destroyed in

---

[116]*Ibid.*

[117]*Ibid.*

[118]*Ibid.*

[119]Organization for Economic Cooperation and Development (OECD) and OECDiLibrary, *Op.cit.*

[120]Huang, Jikun, Jun Yang and Scott Rozelle, *Op.cit.*

[121]*Ibid.*

the 1850s Taiping rebellion, exposing major segments of the population to flooding, droughts, pestilence, and other factors that cause famines.[122]

China is catching up fast with the developed economies in agricultural technologies. The gap in Gross expenditures in research and development (GERD) as a percentage of GDP with the OECD average has been decreasing compared to the early 2000s, when the GDP share of gross R&D expenditure was less than 50% of the OECD average.[123]

It is setting up large-scale parks to host start-ups and other tech firms. China constructed National Agricultural Science and Technology Parks (NASTPs) agricultural technology parks to showcase new tech and promote cooperation between agriculture and other industries, form innovation hubs, create an entrepreneurial chain to augment the transformation and incubation of agricultural sciences and technologies, merge non-profit and profit services, special and comprehensive services with each park made up of a core area, demonstration area and an extended zone to implement "government guidance, enterprise operation, intermediary participation, farmers benefit".[124] The establishment of such parks was also endorsed and encouraged by the state. The state's 2017 No. 1 Document encouraged progress in setting up National Agricultural Science and Technology Parks (NASTPs), agricultural technology parks to showcase innovative technologies, R&D outputs, human resources (HR) training, and new business plans.[125]

Biotechnology for the Future

For the next leap in technology, the Chinese authorities have singled out biotech as a priority industry since the early days of the 1970s. By the late 1970s, China started genetic engineering research.[126] Consistent state funding for biotech development began with the National High Technology Development ("863") Program in March 1986, and by the mid-1980s, transgenic crops were cultivated, virus-resistant tobacco was

---

[122] Talal Abu-Ghazaleh Global (TAG, Global), *Op.cit.*

[123] Organization for Economic Cooperation and Development (OECD) and OECDiLibrary, *Op.cit.*

[124] *Ibid.*

[125] *Ibid.*

[126] Organization for Economic Cooperation and Development (OECD) and OECDiLibrary, *Op.cit.*

commercialized in early 1990, and genetically modified (GM) papaya and cotton were sold since 1997.[127]

Sophisticated experimental test-beds include systems for space-breeding new rice, wheat, cotton, green pepper, and tomato strains, livestock embryo transfer/splitting, gender determination in in-vitro fertilization, test-tubing sheep and cattle, cloning sheep/cattle, genetically-modifying cattle, and the creation of new high-quality aquatic species with nuclear transplantation and transgenic technologies.[128]

In some areas, China is leading the world in biotech products. In 2014, China became the 6th biggest producer of transgenic crops globally, mainly due to the 3.9 million hectares of insect-resistant Bt cotton produced by 7.1 million farmers, which had been commercially sold since 1997, and such transgenic cotton eventually made up about 80% of total Chinese cotton production.[129] Since the 13th Five-Year Plan (FYP) for National Science and Technology Innovation 2016, China has been hastening the breeding of Genetically Modified Organisms (GMOs) as an R&D priority, encouraging the commercialization of key products like new-gen Bt cotton, Bt corn, and herbicide-tolerant soybeans, and instituting biosafety evaluation to ensure the safety of genetically engineered products.[130]

In addition to greater productivity, Bt cotton has reduced environmental impact and input costs through a reduced demand for pesticides, e.g., Hebei Bt cotton resulted in a 55% decrease in pesticide use, while the proliferation of Bt cotton has reduced bollworm infestation on crops within its proximity.[131]

Due to the increasing cultivation of genetically-modified foods and concerns about bioethics, China has set up its own institutions to tackle this issue. It set up a biosafety regulatory system in the mid-1990s and augmented it in the early 2000s.[132] The National Biosafety Committee (made up of 44 specialists drawn from different Chinese ministries, research institutions, and tertiary institutions) put in place guidelines for

---

[127] *Ibid.*

[128] *Ibid.*

[129] *Ibid.*

[130] *Ibid.*

[131] *Ibid.*

[132] *Ibid.*

biosafety assessment (environmental and food safety).[133] There are constant tweaks to its biosafety regulations on transgenic organisms, e.g., the 2016 amendments eliminated timelines for approvals, lengthened the National Biosafety Committee's term from 3 to 5 years, and reiterated that entities engaging in GMO research/experiments are responsible for safety management.[134]

After examining two macro chapters with an overview of agricultural development in China, the next chapter is a specific case study focused on mushroom cultivation in China. Mushrooms appear to be growing in popularity for both cultivators and consumers in China (and indeed the world). The chapter details the contemporary historical survey of mushroom cultivation technologies and know-how in the People's Republic of China (PRC). While it does not pretend to be comprehensive, the major milestones of mushroom cultivation in contemporary China are outlined.

Bibliography

CGTN, "China's smart solar-mushroom plant generates power, boosts production" dated 14 July 2022 in CGTN [downloaded on 14 July 2022], available at https://news.cgtn.com/news/2022-07-14/China-s-smart-solar-mushroom-plant-generates-power-boosts-production--1bEYQM7J1ks/index.html.

CGTN, "China's smart solar-mushroom plant generates power, boosts production" dated 14 Jul 2022 in The Standard Hong Kong [downloaded on 14 July 2022], available at https://www.thestandard.com.hk/breaking-news/section/3/192305/China's-smart-solar-mushroom-plant-generates-power,-boosts-production.

Ge, Siyi, "China Trip Unveils Morel Cultivation Mysteries" dated 13 March 2017 in PennState Collge of Agricultural Sciences Department of Plant Pathology and Environmental Microbiology website [downloaded on 13 March 2017], available at https://plantpath.psu.edu/news/china-trip-unveils-morel-cultivation-mysteries.

Global Times, "China's mushroom tech helps alleviate poverty in South Pacific, Africa and Latin America: delegate to 20th CPC National Congress" dated 16 October 2022 [downloaded on 16 Oct 2022], available at https://www.globaltimes.cn/page/202210/1277220.shtml.

---

[133] *Ibid.*

[134] *Ibid.*

Nandy, Anirban and Fulrida Ekka, "How India could become a 'mushroom superpower'" dated 17 June 2022 in BBC [downloaded on 17 June 2022], available at https://www.bbc.com/news/business-61420016.

Robinson, Nick, "The History of Mushroom Farming" dated 21 Feb 2023 R&R Cultivation Twin Cities Minnesota [downloaded on 21 Feb 2023], available at https://rrcultivation.com/blogs/mn/the-history-of-mushroom-farming.

Long, Yun and Bi Weizi (with contributions from Kong Youqiong), "Mushroom Expert Makes His Mark in Yunnan" dated 23 November 2023 in Science and Technology Daily [downloaded 23 November 2023], available at http://www.stdaily.com/English/Service/202311/d468a545f7584d7781568201be6c7b97.shtml.

Shen, Weiduo, "Xiconomics in Practice: Xi-endorsed Juncao technology improves livelihoods across globe, embodies China's commitment for greater good of humanity" dated 12 October 2023 in Global Times (GT) [downloaded on 12 October 2023], available at https://www.globaltimes.cn/page/202310/1299737.shtml.

Xinhua, "Ethiopia teams up with China on mushroom project" dated 19 December 2019 in China Daily [downloaded on 19 December 2019], available at https://global.chinadaily.com.cn/a/201912/19/WS5dfaec4ea310cf3e3557f3e0.html.

Chapter 5

Case Study of Mushroom Cultivation Technologies: Domestic Development and Exporting Technologies Overseas

Tai Wei Lim

Abstract

This chapter on China briefly surveys the history of technological use in Chinese agriculture and food sectors. It adopts area studies and political economic angles to examine this topical matter with a focus on policy analysis related to the utilization of Industry 4.0 technologies.

Introduction

Some researchers argue that mushrooms are the "most well-known and recorded edible forest product" on Earth, but, at the same time, even though mushrooms have been grown since ancient times, the scalability of mushroom-growing was often circumscribed and restricted to a handful of optimal climates, regions, and seasonal variations.[1] As early as the time

---

[1] Adusumalli, Harshini Priya, Nur Mohammad Ali Chisty, Rahman, Mahofuzur, Hossain, Shakawat Hossain and Pasupuleti, Mahesh Babu, "Bio-innovation and technological digitization of mushroom cultivation and marketing for rural development" dated 2022 in Academy of Marketing Studies Journal, 26(S3) [downloaded on 31 Dec 2022], available at https://www.abacademies.org/articles/bioinnovation-and-technological-digitization-

of the first human civilizations, the benefits of mushrooms had already been realized. In the ancient world, the Egyptian Pharaohs considered mushrooms to be a culinary delight and a nutritious source of medicinal ingredients and health supplements, with a wide spectrum of minerals and natural phytochemicals embedded within mushrooms that contribute to health benefits.[2]

Other sources indicated that China was, in fact, the progenitor of mushroom cultivation. The first curated evidence of humans cultivating mushrooms was discovered in China when peasants started cultivating shiitake mushrooms for various reasons (e.g., making animal feed for husbandry) over a millennium ago, but growing mushrooms specifically for food started in 600 AD.[3] The shiitake mushroom (*Lentinula edodes*) was apparently first domesticated by Chinese cultivators, and they instructed Japanese peasants on the methods of growing the shiitake. The Japanese eventually improved those methods and expanded their trade in East Asia.[4]

Mushrooms were nicknamed the "food of the gods" by the Romans, and, to the Greeks, they enhanced the strength of warriors for conducting warfare, and therefore, these folklores have made mushrooms "mystical, cultural, traditional, and legendary".[5] While they may be the stuff of legends in the past, in the modern era, a mushroom is scientifically defined as the "fleshy, spore-bearing fruiting body of a fungus that is normally grown above ground on soil or on its feeding supply, with the majority of mushroom production occurring in forest environments" and there are about 14,000 of the 1.5 million recorded fungal species which have fruiting bodies sufficiently sizable to be labeled as "mushrooms".[6]

In the mid-20th century, new technologies and techniques emerged in the mushroom-growing industry with synthetic compost and climate-controlled environments facilitating full-year cultivation activities,

---

of-mushroom-cultivation-and-marketing-for-rural-development-14654.html, pp. 1–7. {unpaginated online version was reference for this writing}.

[2] *Ibid.*

[3] Robinson, Nick, "The History of Mushroom Farming" dated 21 Feb 2023 R&R Cultivation Twin Cities Minnesota [downloaded on 21 Feb 2023], available at https://rrcultivation.com/blogs/mn/the-history-of-mushroom-farming.

[4] Adusumalli, Harshini Priya, Nur Mohammad Ali Chisty, Rahman, Mahofuzur, Hossain, Shakawat Hossain and Pasupuleti, Mahesh Babu, *Op.cit.*

[5] *Ibid.*

[6] *Ibid.*

bumping up production and reducing expenditure that made it possible to supply supermarkets and restaurants globally, thereby turning it into a staple food source.[7] Synthetic logs constructed out of sawdust and millet and wheat bran can churn out 3–4 times as many mushrooms as natural timber logs in 1/10th of the time while temperature, humidity, light, and log moisture may all be climatically controlled in indoor rooms to optimize harvests; and, for specific genres like shiitake, synthetic logs work better than natural logs facilitating year-round supply, shorter yields, and more compressed crop cycles.[8]

Additional nutrients are added for optimal shiitake growth. Sawdust is the typical base element in shiitake substrate contents, along with ingredients like straw, corncobs, and/or other starch-based supplements (10–60% of the dry weight) like wheat, rice bran, calcium carbonate (CaCO3), gypsum, and table sugar to provide nutrients to facilitate mushroom growth.[9]

From 1969 to 2009, global mushroom output accelerated by a factor of 10 times with the Food and Agriculture Organization (FAO) of the United Nations (UN) detailing sizable expansion in China, the US, the Netherlands, India, and Vietnam (especially with China expanding at a 10% yearly rate since the early 1980s to 2022).[10] While this technological revolution was happening in the world, China experienced large-scale ideological campaigns. Before the 1978 economic reforms in the People's Republic of China (PRC), the Chinese mushroom-growing economy consisted of a commune system that coordinated all major economic operations in rice/wheat cultivation and retailing, though the state considered mushroom cultivation to be a minor farm activity.[11] Thus, its development was de-privileged.

However, the high costs of some traditional mushrooms like white jelly fungus and shiitake motivated many farmers to take up wild mushroom gathering (nicknamed "treasure seeking") and innovate new mushroom-growing methodologies for exporting to overseas ethnic Chinese diasporic customers.[12] They have the same understanding of the

---

[7] Robinson, Nick, *Op.cit.*

[8] Adusumalli, Harshini Priya, Nur Mohammad Ali Chisty, Rahman, Mahofuzur, Hossain, Shakawat Hossain and Pasupuleti, Mahesh Babu, *Op.cit.*

[9] *Ibid.*

[10] *Ibid.*

[11] *Ibid.*

[12] *Ibid.*

benefits of edible Chinese fungus/mushrooms, and, for the state-owned enterprises in China, it was a valuable opportunity to earn valuable foreign reserves for the foreign trade department during the pre-reform days.[13] The high-value-added products of white jelly fungus and shiitake, used mainly for Chinese medicine and exported by the Chinese government until the 1970s to elicit precious foreign currency, motivated the state (especially its foreign trade department) to finance and set up research institutions to study mushroom species and their associated resources.[14]

Domestically, a handful of main state-owned enterprises farmed mushrooms as an ancillary vegetable-related secondary product starting in Beijing and Shanghai in the early 1980s, while, in other parts of the country, many rural households cultivated mushrooms, particularly in a few geographical locations in warm and humid southern China.[15] However, the mushroom-growing regions soon shifted in locations beyond the traditional south, made possible by technology solutions and economic reasons.

Approximately, 25 million individual are currently working in the mushroom growing and processing industries and the traditional hub of mushroom cultivation remains in semi-tropical and humid southeast China but, from the 1990s onwards, mushroom production has relocated northwards where personnel costs are lower, raw materials are more affordable and abundant (e.g., north-eastern China timber and forest land or central China which generate large amounts of agricultural waste) and proximate to large northern urban consumer markets like Beijing and Tianjin.[16]

Higher labor and material costs in southern China (which had good natural conditions for growing mushrooms) pushed enterprising farmers to move to cheaper locations to cultivate mushrooms, resulting in substantial penetrative technological diffusion throughout the country.[17] The Chinese Edible Fungi Association calculated that there were 15 million rural farmers producing and processing edible fungi in the early 2000s, and some of them like Chinese peasants Dai Weihao, Yao Shuxian, and

---

[13] *Ibid.*

[14] *Ibid.*

[15] *Ibid.*

[16] *Ibid.*

[17] *Ibid.*

Pan Zhaowan from Gutian County (Ningde City Fujian province) grew white jelly fungus in glass bottles with woody sawdust before turning to using plastic bags and their successes in cultivation were publicized as heroic pioneers by the local community and national mass media.[18]

Successful southern mushroom farmers served as inspiration for other regions in China. As a long-running economic industry in southern China, many agricultural institutions like the Shanghai Academy of Agriculture's Edible Mushroom Research Institute and South-Central Agricultural University established research teams for mushroom cultivation, and, in the early 1980s, the Beijing Municipal Government wanted to farm mushrooms as an alternative to vegetables.[19] Thus, mushroom cultivation began to head north.

After the economic reforms in 1978 and the US–China rapprochement, China obtained access to the advanced technologies of the West and the international community. With such technological development, Chinese output began to increase tremendously. Some state-advocated/subsidized integrated enterprises started coordinating with proximate farms in their locations to lower costs, augment productivity, and market share by assigning some responsibilities to farmers who are provided with substrate bags/technology and then delivering the outputs to the integrated enterprises, which ensures quality control and food safety for the products.[20] The government, universities, and provincial research institutions were all important in discovering new mushroom species and growing them, while local institutes made available different breeds and conserved cultures for the bigger mushroom farms, and some small private institutions produced mushrooms to sell to small-time rural mushroom farmers.[21]

This rationalization process was coupled with centralized social mobilization that made large-scale mushroom cultivation possible. In China, local village committees put together trainings, workshops, and seminars featuring successful mushroom producers or specialists, as well as organized activities that were well-received by farmers.[22] The China Edible Fungi Commerce Network released data in 2003 that indicated edible fungi consumption per capita in Shenzhen, Shanghai, and Beijing

---

[18] *Ibid.*

[19] *Ibid.*

[20] *Ibid.*

[21] *Ibid.*

[22] *Ibid.*

hit 6 kg, surpassing that of the US, Eurpope, and Japan, and this increased to more than 10 kg annually in 2008.[23]

Globally, R&D researchers are looking at the utilization of robotics and artificial intelligence (A.I.) to maximize mushroom production, and some are examining ways to utilize recyclable waste materials and other sustainable practices in mushroom cultivation.[24] One of the new contemporary growth areas in agro-technological innovation is in the area of mushroom cultivation. China has come up with an innovative approach that merged solar power generation and smart manufacturing to increase local electricity generation and improve smart mushroom cultivation in Qingyuan County's Lishui City in eastern Zhejiang Province.[25]

Here, solar panels placed on roofs of the 12,000 sq. metre mushroom plant facility can churn out 3.05 million kilowatt hours of power/electricity, saving 280,000 yuan (approx. US$ 41,600) in electricity costs annually, with Li Zhijun (State Grid Zhejiang Qingyuan county technician) articulating: "For the next step, we will further the electrification transformation of the mushroom plant and build a new solar-mushroom project by 2025, with a capacity of 70 megawatts, in a bid to fuel the high-quality development of the edible fungi industry in Qingyuan."[26]

In these solar-powered mushroom plants, the plant personnel are hard at work producing mushroom spawns as a substrate for edible fungi cultivation in an industrial environment, a contrast from the past practice of mushroom spawn production that centered on distribution by local farmers, which incurred large labor costs, low production volume, lower quality, and yields.[27]

Several bio-innovations in the late 1970s and early 1980s catalyzed the accelerated development of growing shiitake and other mushrooms on natural logs extracted from hardwood trees in the autumn season after the leaves have dropped from the trees and inoculated with shiitake

---

[23] *Ibid.*

[24] Robinson, Nick, *Op.cit.*

[25] CGTN, "China's smart solar-mushroom plant generates power, boosts production" dated 14 July 2022 in CGTN [downloaded on 14 July 2022], available at https://news.cgtn.com/news/2022-07-14/China-s-smart-solar-mushroom-plant-generates-power-boosts-production--1bEYQM7J1ks/index.html.

[26] *Ibid.*

[27] *Ibid.*

spawn 15–30 days later.[28] The cut trees are left as they are in the winter season and then cut up into 1 m-long planks before they are inoculated with spawn in the form of wooden plugs or sawdust, and finally, holes are drilled into the logs with specifications that fit the spawn's diameter and length.[29] This conventional methodology needed low wood and labor productivity; thus, the methodology was not substantially changed or improved over a long time.[30]

Solar technologies have also been installed to harness sunrays to power the mushroom-growing processes. The 30-million yuan (approx. US$ 4.45 million) digital mushroom production line armed with photovoltaic system in Qingyuan County's Lishui City in eastern China's Zhejiang Province experienced significant increase in mushroom spawn production and Ye Gao (leader of the mushroom plant) revealed that, from the start of its operations, the daily output of the plant has gone up substantially: "Meanwhile, more than 30 percent of the labor cost has been cut and the yield rate reached almost over 99.9%."[31]

Chinese use of technologies for mushroom cultivation has contributed to the country becoming a mushroom-growing superpower, making up 75% of the global share, while another equally large, fast-developing country, India, only accounts for 2% of the world's mushroom production, despite the latter having better natural conditions for mushroom-growing.[32]

While advanced technologies do hold positive promises for the future, there is a limit to mushroom cultivation expansion made possible by tech enhancements. The lesson of an advanced economy may be instructive here. A former leader in biotech and shiitake cultivation, high Japanese personnel costs resulted in losing market shares to farmers in China. Indonesia, India, and Vietnam, as there are still many processes that are labor-intensive, favoring Global South farmers who also receive support

---

[28] Adusumalli, Harshini Priya, Nur Mohammad Ali Chisty, Rahman, Mahofuzur, Hossain, Shakawat Hossain and Pasupuleti, Mahesh Babu, *Op.cit.*

[29] *Ibid.*

[30] *Ibid.*

[31] CGTN, *Op.cit.*

[32] Nandy, Anirban and Fulrida Ekka, "How India could become a 'mushroom superpower'" dated 17 June 2022 in BBC [downloaded on 17 June 2022], available at https://www.bbc.com/news/business-61420016.

from the Food and Agriculture Organization.[33] Above 25 million farmers in China are involved in the harvesting, cultivation, processing, and retailing of mushrooms.[34]

In the case of China, its natural mushroom-growing region is Yunnan. South-western Chinese Yunnan province is a picturesque location with multi-ethnic communities that also hosts a wide spectrum of mushroom species that thrive in the unique mountainous environment.[35] Even now, a handpicked collection of mushrooms continues. Edible mushrooms were formerly extracted in the wild, as mushrooms were challenging to domesticate and cultivate in the lab; thus, even in the modern/contemporary world, some food gatherers in southern Asia and other developing countries still collect their mushrooms from natural forests.[36] But, as the following section indicates, the use of advanced technologies can now make mushroom-growing possible in any part of China.

Cultivation of Morel Fungi

Within the fungi industry, perhaps one of the most value-added products is morel fungi. There are three clades of *Morchella spp.* and more than 30 species: *rufobrunnae clade, esculenta clade*, and *elata clade*. Chinese experts have zoomed in on artificially-cultivated species like *M. importuna, M. sextelata, M. septimelata*, and *M. esculenta* with *M. importuna* the most popularly-grown species in China (more than 95%) due to its high productivity and growth stability.[37]

---

[33] Adusumalli, Harshini Priya, Nur Mohammad Ali Chisty, Rahman, Mahofuzur, Hossain, Shakawat Hossain and Pasupuleti, Mahesh Babu, *Op.cit.*

[34] *Ibid.*

[35] Long, Yun and Bi Weizi (with contributions from Kong Youqiong), "Mushroom Expert Makes His Mark in Yunnan" dated 23 November 2023 in Science and Technology Daily [downloaded 23 November 2023], available at http://www.stdaily.com/English/Service/202311/d468a545f7584d7781568201be6c7b97.shtml.

[36] Adusumalli, Harshini Priya, Nur Mohammad Ali Chisty, Rahman, Mahofuzur, Hossain, Shakawat Hossain and Pasupuleti, Mahesh Babu, *Op.cit.*

[37] Ge, Siyi, "China Trip Unveils Morel Cultivation Mysteries" dated 13 March 2017 in PennState Collge of Agricultural Sciences Department of Plant Pathology and Environmental Microbiology website [downloaded on 13 March 2017], available at https://plantpath.psu.edu/news/china-trip-unveils-morel-cultivation-mysteries.

Historically, Chinese scientists carried out morel cultivation research since 1980 and authored successful production reports on best practices by the likes of Douxi Zhu (Head, Mianyang Edible Fungi Research Institute of Sichuan), whose institute initiated morel artificial cultivation in 1985 and achieved their inaugural successful morel fruiting bodies in 1992.[38]

By 2012, their methodologies were applied in farms nationwide, and according to the statistical data provided by the China Edible Fungi Industry Chapter, the total morel-growing lands in China are 23,400 acres, up from only 1000 acres in 2011, and almost four times the land size in 2014 in more than 20 provinces.[39] Sichuan province makes up 44% of China's overall morel production area, and the second biggest morel growing province is Hubei (including its Songzi City).[40]

The success of Zhu's innovation techniques was partly due to accidental fate. Dr. Douxi Zhu's technique in moral-cultivation involved the application of nutrient bags to the crops and, in 2000, Zhu hired farmers to assist with spawn bags (mycelium to inoculate the soil) logistics, but they inadvertently left a few bags on the soil surface and, and 10 days later, Zhu discovered white, powdery substances around the bags with openings facing downward.[41] When fruiting bodies emerged, Zhu discovered more morels in the area where the forgotten bags were left, prompting him to retrace and recreate the steps.[42] Just like that, his indigenous methodology was born.

US universities like PennState sent delegations (led by Dr. John Pecchia and his Master of Science M.S. student Siyi Ge) to the Chinese Morel Industry Conference and the Development of Global Valuable Edible Fungi Forum held on 10–12 March 2017, where more than 400 individuals participated (with farmers, researchers, investors, and morel fans).[43] The morel industry conference was an apt platform for Siyi Ge and Dr. Pecchia to observe how agriculturalists cultivate morels (*Morchella spp.*) in China, utilizing updated research and methodologies,

---

[38] *Ibid.*

[39] *Ibid.*

[40] *Ibid.*

[41] *Ibid.*

[42] *Ibid.*

[43] *Ibid.*

a value-added activity for Pecchia's team, given that there is no morel farming industry in the US.[44]

US observing the Chinese morel industry and its best practices is an ironic development, given that Ronald D. Ower first artificially cultivated them in the US in 1982 and secured the earliest patent for this technique in 1985 (No. 4594809).[45] While a Michigan State University morel farm in 2005 initiated commercially-cultivated morels, it ceased its operation in 2008 because of technical difficulties, bringing down the entire industry in the US with it.[46] Ronald Ower may have indicated the additional nutrient source concept in 1982, but did not realize its significance in morel cultivation, and also did not fully appreciate the scientific fact that *Morchella spp.*'s sexual development needed a comparatively nutrient-poor environment. This was something that the Chinese researchers realized from their own research later on.[47]

Nutrient-deficient soil cannot sustain new mycelia growth without adding more nutrients for crop cultivation, so Chinese cultivators stuffed bags with wheat, sawdust, corn, etc., and positioned the bags on the soil surface several days after spawning to boost productivity.[48] They then take away the bags to trigger the *Morchella spp.*'s sexual development in regional farms located south of the Changjiang River (Hubei province), while farmers spawn from mid-October to mid-November in regions north of the Changjiang River, starting from mid-September.[49]

There remains, however, much room for further improvements. Plastic nutrient bags and greenhouses can be very prohibitive in costs and so spawn sellers, so training agencies and bidders benefit more than the local peasants who bear the financial burden in buying the materials and land that cost approximately 10,000 CNY (US$1,430) per acre.[50] Thus, farmers are pressured to harvest a minimum of 67 kg of fresh morels per acre to maintain profitability, something difficult to achieve and sustain for a substantial number of farmers.[51]

---

[44] *Ibid.*

[45] *Ibid.*

[46] *Ibid.*

[47] *Ibid.*

[48] *Ibid.*

[49] *Ibid.*

[50] *Ibid.*

[51] *Ibid.*

While the morel industry has seen experts from advanced economies like American researchers observing Chinese successes, other Chinese mushroom/fungi products, like the Juncao grass, are marketed to developing economies instead. The following Juncao case study also contrasts with the morel case study, as the former generates low-cost feed for animals while feeding humans at the same time; morels are one of the most valuable, expensive, and value-added products in the fungi world.

Juncao Technology

Conventionally, most mushrooms are still cultivated using compost and climate-controlled environments, but countries like China are trying to evolve new methodologies to disrupt the industry.[52] Enter the Juncao technology. The name "Juncao" is a combination of the words meaning "mushroom" and "grass" that connote cultivating edible mushrooms as livestock feed or even as an environmental green barrier to arrest desertification.[53]

In 2001, the Fujian Provincial authorities initiated the Juncao and upland rice technology project in the Eastern Highlands, starting with 7 technical training courses to 623 people, cutting down poverty and improving food security, according to a report on the collaboration's outcomes published by the Fujian local government in May 2022.[54] Local governments gave incentives and support to successful farmers in reducing or absorbing any potential losses in their ventures, becoming a buffer against widespread losses if any of the crucial stages of mushroom cultivation fail.[55] The local governments also established publicly-accessible demonstration places to showcase the effectiveness of certain cultivation techniques directly to the farmers while taking questions from them.[56]

---

[52] Robinson, Nick, *Op.cit.*

[53] Global Times, "China's mushroom tech helps alleviate poverty in South Pacific, Africa and Latin America: delegate to 20th CPC National Congress" dated 16 October 2022 [downloaded on 16 Oct 2022], available at https://www.globaltimes.cn/page/202210/1277220.shtml.

[54] *Ibid.*

[55] Adusumalli, Harshini Priya, Nur Mohammad Ali Chisty, Rahman, Mahofuzur, Hossain, Shakawat Hossain and Pasupuleti, Mahesh Babu, *Op.cit.*

[56] *Ibid.*

Municipal governments showcase the best quality products, while the authorities in Gutian and Qingyuan funded mushroom museums for the advocacy of mushroom cultivation culture, curated their histories, and featured local products.[57] Local governments also host activities to market local mushrooms, their equipment, and other tech innovations, attracting businesses nationwide, and they also often finance celebratory and cultural events like music/artistic performances, cooking demonstrations, competitions, marketized promotion, nutrition instructional workshops, conferences, and seminars to create public awareness.[58] Seasonal agricultural product expos in major cities (e.g., Shanghai, Beijing, Guangzhou, Hangzhou) have become very important marketing outlets for facilitating: (1) direct retailing from rural farmers to urban consumers; (2) organizing meetings and interactions between producers and businesspeople; and (3) promotion of mushroom cuisine demonstrations and cooking knowledge to many consumers.[59]

Such environments are different from the climates, temperatures, and terrains of mushroom farms in other world regions. Mushrooms were handpicked from forests by the earliest hunter-gatherers, but gradually became cultivated over time to become the globe's biggest agricultural sector.[60] Until 3–4 centuries ago, harvesting mushrooms mainly through handpicking remained the case in some world regions. Afterwards, Europeans began cultivating mushrooms. For example, during the 17th century, mushroom-growing proliferated amongst French gardeners who were pioneers in cultivating cave mushrooms in caves, which are considered the optimal environment for growing mushrooms, a technique that soon spread Europe-wide as a standardized template for cultivating mushrooms that lasted until the mid-19th century.[61] Therefore, this shows that there is space for alternative techniques that can suit different parts of the world with a spectrum of terrains, climates, and weather systems.

Soil conditions are indeed very important. For example, far more than crops like Agaricus bisporus (the best-selling edible mushroom in the US), shiitake, and oyster mushrooms, cultivating fungi like morels is mostly reliant on climatic and soil conditions which can vary substantially

---

[57] *Ibid.*

[58] *Ibid.*

[59] *Ibid.*

[60] *Ibid.*

[61] Robinson, Nick, *Op.cit.*

in various geographical sites (the cultivation of mushroom/fungi crops reliant on both natural and artificially-induced environmental factors is known as a bionic process).[62]

Wood and timber are also important substrate materials. The raw materials utilized as substrate for mushrooms range from hardwoods to other kinds of woods, with the percentage of agricultural residues and wastes increasing with time.[63] Subsequently, with technological improvements, synthetic wood logs made of wood dust (sometimes with millet and wheat bran too) were then used in place of natural wood logs in China and this augmented production by a factor of 15–20 times and reduced the cultivation period by 150 days, while cotton residuals doubled the product output.[64]

In the contemporary era, mushroom cultivation is incrementally dependent on Chinese agricultural and wood wastes, something increasingly important due to the increasing cost factor of wood/timber (which has also catalyzed rapid expansion of mushroom-growing substrate production in a few locations).[65] The contesting economic demand for wood as a raw material has motivated the stakeholders to set up alternative forest production and management methodologies for industries like the mushroom cultivation industry in China, to create specialized mushroom forests for research on specific mushroom species and silviculture, including those that grow well in the forest's understory and floor.[66]

In the case of Juncao, it has developed its own patented methodology. Cultivating Juncao mushrooms is multi-staged. State-owned *Global Times* journalists covering this process noted the following procedures: "[involving] strain cultivation, Juncao grass grinding, mixing of substrates, bagging, sterilization, inoculation, mycelium cultivation, and mushroom harvesting. Mycelium cultivation in the bags before mushroom harvesting is a critical and time-consuming step with some technical intricacies. In the current model in Fiji, the China-Aided Fiji Juncao Technology Demonstration Center provides farmers with pre-cultivated bags, and farmers only need to complete the final step, which is mushroom harvesting. [The *Global Times*

---

[62] Ge, Siyi, *Op.cit.*

[63] Adusumalli, Harshini Priya, Nur Mohammad Ali Chisty, Rahman, Mahofuzur, Hossain, Shakawat Hossain and Pasupuleti, Mahesh Babu, *Op.cit.*

[64] *Ibid.*

[65] *Ibid.*

[66] *Ibid.*

reporters] observed Juncao grinding machines, bagging machines, and simple atmospheric sterilization stoves provided by China in the college's open area. In a greenhouse converted from a shipping container, the Juncao mushrooms are growing well."[67]

Knowledge Transfers and Instructions

Lin Xingsheng (leader of the Juncao specialist group) informed the *Global Times (GT)* that: "[Juncao technology covers a wide range of applications and offers extensive opportunities for the development of the Juncao industry chain in the island nation.] Juncao can serve as a substrate for cultivating edible and medicinal fungi. Additionally, it can be used as an effective feed for livestock like cattle and sheep, addressing the issue of feed shortage during the dry season in the drought-prone areas of western Fiji and promoting local livestock development. [Juncao exhibits rapid growth and a well-developed root system, making it useful for addressing local issues such as soil erosion and soil salinity in addition to its other benefits.] But we hope that local farmers can independently carry out the entire process, from bag production to mushroom harvesting [noting that Votualevu College, a vocational and technical institution where students learn specific skills and subsequently enter the workforce in Nadi, is a pilot point for the training]."[68]

Unlike other industries and agricultural production, mushroom cultivation skills are quite foundational and relatively easy to master directly with only minor support needed from the state, and much of the knowledge transfer (and trade secrets) is informally extracted from relatives or neighbours (and other social networks).[69] Practical farming knowledge was mainly formulated at the grassroots by farmers and other stakeholders who may not necessarily be qualified professionals, and, in the late 1970s and early 1980s, farmers and private research entities started experimenting with cultivating mushrooms in glass bottles and later plastic bags,

---

[67] Shen, Weiduo, "Xiconomics in Practice: Xi-endorsed Juncao technology improves livelihoods across globe, embodies China's commitment for greater good of humanity" dated 12 October 2023 in Global Times (GT) [downloaded on 12 October 2023], available at https://www.globaltimes.cn/page/202310/1299737.shtml.

[68] *Ibid.*

[69] Adusumalli, Harshini Priya, Nur Mohammad Ali Chisty, Rahman, Mahofuzur, Hossain, Shakawat Hossain and Pasupuleti, Mahesh Babu, *Op.cit.*

while transiting from indoor to outdoor farms and from woodlands to wastes/agricultural residuals.[70] It was during this period that experimental data indicated that every 1000 synthetic log bags and cotton residuals can harvest 75 kg of mushroom while wood only comes up to 25–35 kg, before the era of technological advancements like disease prevention, specialized division of labor using new machines, and raw ingredients improved yields further as the farmers studied the basic research data generated by state organizations.[71] But, as the scale of mushroom-growing began to expand, and the land under cultivation grew in size, necessitating professionalization that required formal instructions and institutionalized ways of knowledge transfer.

Praveen Chand, an instructor at the Votualevu College, told the state-owned *Global Times (GT)*: "We started three years back, but from this year we have gone into commercial scale. At the moment we are only for our school but we also are sharing this knowledge with the community too. The money that the school normally gets from the mushrooms is reinvested into the teaching and learning of our students who are in the vocational agriculture industry. And the students have shown a lot of interest and they are quite keen in terms of learning. [China's training program doesn't only target young people but also includes other professional technicians. Praveen Chand himself had previously visited China for short-term technical training. To date, the Chinese expert group has organized a total of 40 training sessions across various locations in Fiji, benefiting nearly 2,300 individuals. In addition, 54 Fijian agricultural officials and technical personnel were selected to receive training at the Fujian Agriculture and Forestry University in China.]"[72]

Taufa Tupou, a Votualevu College student, researched and studied mushroom cultivation for a year, and she told the state-owned *Global Times (GT)* that she is cultivating Juncao at her own facility and plans to go fully into the industry as her career.[73] Vatimi Rayalu, Fiji Minister of Agriculture and Waterways, showed his gratitude to the Chinese authority for sharing the Chinese experience in cultivating Juncao in March 2023, when the China-Pacific Island Countries Juncao Technology Demonstration Center was officially inaugurated in Fiji, on

---

[70] *Ibid.*

[71] *Ibid.*

[72] Shen, Weiduo, *Op.cit.*

[73] *Ibid.*

22 March 2023: "Juncao can be a great solution to soil erosion and desertification, helping to address natural disasters and food crises, becoming an important supplement to traditional farming, and contribute to the empowerment of young women."[74]

Lin Zhanxi (delegate to the 20th National Congress of the Communist Party of China (CPC) and inventor of Juncao technology) said on 16 October 2022 at a press event for the opening of the 20th CPC National Congress at the Great Hall of the People in Beijing, explained:

"Papua New Guinea (PNG) was the first country that received aid from China for Juncao technology [noting that his team tried their best to localize and simplify the technology of breeding fungi with herbaceous plants so that rural residents in PNG could easily master it]. To celebrate the success of Juncao technology demonstration projects, the PNG side held a grand ceremony, with participants including then Prime Minister, nine ministers and over 5,000 people from across the country, who hailed "China Herb," and "Herb China". In order to remember China's help, a minister changed the name of his daughter to "Juncao". Since then, Juncao technology has helped alleviated poverty in South Pacific nations, Africa and Latin America and changed the destiny of those people. Juncao technology has been promoted in 18 languages. Over 12,000 people have been trained both at home and abroad."[75]

China is exporting its advanced technologies overseas. Juncao is now patented and it is touted as easy to manage and profitable, and, according to the Chinese Foreign Ministry, the Juncao technology has been marketed in 15 languages and more than 10,000 individuals were instructed worldwide in cultivating the grass, including in instructional and demo hubs in 13 Asian, African, and Pacific states.[76]

Besides Juncao, there were also large-scale transfers of knowledge for Chinese indigenous mushrooms as well. Before the late 1980s, mushroom cultivation was an activity with a fast turnover and successful wealthy farmers and villages inspired other farmers in other areas to pick up the technology or even hired expert or experienced farmers as instructors or technicians to transplant mushroom-growing knowledge to a new area/region to lift up revenues there, incentivizing the experts with a good pay,

---

[74] *Ibid.*

[75] Global Times, *Op.cit.*

[76] Shen, Weiduo, *Op.cit.*

local appointments, land parcel offers and other perks.[77] This process diffused mushroom cultivation technical knowledge far and wide in the late 1980s and early 1990s, and after acquiring mushroom knowledge, some learner-students worked on mushroom farms instead of looking for salaried employment in other sectors. Since the mid-2000s, hundreds of households rented land and established farms in mushroom-growing regions like Gutian, or went back to their hometowns for the same purpose.[78]

[Analogously, large-scale transfers of mushroom cultivation techniques have occurred in recent world history as well. In the US, for example, mushroom farming took off in the early 20th century when the inaugural US commercial mushroom farm was set up in 1896 by W. Robinson in Pennsylvania but the boom only came in the 1920s when European immigrants transferred their knowledge and expertise to the New World.[79] In those days, indigenization and localization of techniques was the name of the game. When spawn was imported from Europe to the US, it was frequently considered undependable and costly, leading to the establishment of the American spawn industry in the early 20th century that substantially upgraded the quality and availability of spawn for mushroom growers.[80]]

Tang Wenhong (vice-chair of the China International Development Cooperation Agency) noted in March 2023 that the Juncao technology project demo center in Fiji set up with the Fiji authorities in 2014, has turned into the biggest Pacific platform with more than 600 local farmer participating households, more than 1,000 households tapping into the overall cultivation area of more than 7000 mu (466.6 hectare).[81] When the state-owned media, *Global Times*, went to Fiji, they discovered that the grass has contributed to ecological protection, improved local community lifestyles, upgraded female incomes, and encouraged sustainable development for Fiji.[82]

---

[77] Adusumalli, Harshini Priya, Nur Mohammad Ali Chisty, Rahman, Mahofuzur, Hossain, Shakawat Hossain and Pasupuleti, Mahesh Babu, *Op.cit.*

[78] *Ibid.*

[79] Robinson, Nick, *Op.cit.*

[80] *Ibid.*

[81] Shen, Weiduo, *Op.cit.*

[82] *Ibid.*

Neori Tiko, employee at Tadra farm in Nadi, informed the *Global Times (GT)* that his farm began to implement Juncao technology in 2020, in a year of serious drought with withered grass: "This Juncao grass is really, really helpful to us, thank you China. So we got introduced by the "Juncao guys" [Chinese specialists at the Juncao technology demonstration centre in Fiji] from China and started to plant them. From then till now, we use it as feeds for the horses, for mushroom production and are planning for planting more."[83]

In the Central African Republic (CAR), Irene Dombeti (lab technician of the Institute for CAR Agricultural Research) is working on cultivating mushrooms sustainably and technologically after spending 1.5 months in China researching edible and medicinal mushroom technologies. However, after coming back to CAR, she noted that her fellow countrymen/women appeared less enthusiastic about utilizing technologies to cultivate mushrooms and articulated:

> They said that mushroom grows naturally, so they are not interested in mushroom technology. Gradually in seeing what we are doing, they are now starting to adapt to mushroom cultivation because the mushrooms that we are growing here with technology are good mushrooms, which taste good and then it attracts people, they are now much interested. In order to better popularize the Juncao technology, our team is building an 854-square-meter production workshop. We will continue to popularize this technology, train more locals and domesticate the local wild edible and medicinal fungi to help ease the food shortage problem in CAR.[84]

China appears to have developed other patented capabilities in resilient fungi cultivation technologies as well. Chinese experts came up with the Juncao grass-planting technology that is environmentally adaptable to the various external environments and imbued with rich nutrition, which can be processed to grow edible and medicinal fungi and raise poultry and livestock.[85] Chen Kehua brought the technology to the CAR when she

---

[83] *Ibid.*

[84] Tamfu, Arison, "Feature: Chinese Juncao technology brings hope to mushroom growers in Central African Republic" dated 26 July 2022 in Xinhua edited by Huaxia [downloaded on 26 July 2023], available at https://english.news.cn/africa/20220726/0e229669c0bd4a539ff5ce560bb6c2c9/c.html.

[85] *Ibid.*

taught 16 Juncao training courses to 613 trainees, engaged five cooperatives in mushroom cultivation and four cooperatives in feeding farmed animals with Juncao grass, while her team cultivated 3 hectares of Juncao in the process of assisting local peasants to plant 10 hectares in total.[86]

Courses for creating awareness of the nutritional and consumption value of mushrooms are important. It is less crucial in places like China, where mushroom eating has a long history. In contrast, many Chinese think that edible mushrooms/fungi have strong nutritional benefits proven by their time-tested tradition[87] in Chinese Traditional Medicine (TCM) and culinary cuisines. While India had close to 100 types of edible mushrooms, its industry did not take off, as an Indian mushroom specialist, Rouf Hamza Boda, argued that many Indians dislike eating mushrooms and consider them "strange and deadly" because "Not much research has been done on identification of wild mushrooms with respect to their edibility. Lack of awareness as to how beneficial mushrooms are, and the cheapness of mushroom cultivation, are hurdles in popularizing consumption."

This has therefore stifled India's development as a major global player in the mushroom industry.[88] In contrast, China and many other parts of the world embrace mushroom/fungi's health benefits. The health advantages of consuming mushrooms cover wound recovery, enhanced immunity, and the retardation of tumor growth, and, in recent periods, medicinal mushroom experiments carried out on African HIV/AIDS patients sharply increased the public awareness of mushrooms to incredible heights, backed up by positive health experiments and trials.[89]

Fatime Abba Rekya, agronomist and mushroom cultivator, was a Juncao pioneer and aided by Chinse expert to retail the output on the internet and has apparently attained commercial success due to high demand for it after consumers were impressed by the taste: "We are doing our best to gradually settle the orders we have [adding that they have succeeded to train 20 women in mushroom cultivation]. The Juncao technology has contributed to job creation, food security, and generated income for rural

---

[86] *Ibid.*

[87] Adusumalli, Harshini Priya, Nur Mohammad Ali Chisty, Rahman, Mahofuzur, Hossain, Shakawat Hossain and Pasupuleti, Mahesh Babu, *Op.cit.*

[88] Nandy, Anirban and Fulrida Ekka, *Op.cit.*

[89] Adusumalli, Harshini Priya, Nur Mohammad Ali Chisty, Rahman, Mahofuzur, Hossain, Shakawat Hossain and Pasupuleti, Mahesh Babu, *Op.cit.*

communities in CAR, and has played an important role in tackling malnutrition in the country." "It's a business of hope, because there are restaurants and hotels here, and we can dry off and export mushrooms and do a lot of things around it. For young people who have not gone to school, we can create jobs for them, for the women who are at home, we can also create jobs for them."[90]

Ethiopia

Juncao is not the only option from the Chinese mushroom industry in its exchanges with Africa. Another African state working with Chinese mushroom experts is Ethiopia. Ethiopia and China officially inaugurated the "Ethiopia 1550 Mushroom Program" mushroom project that empowers the former with modern advanced scientific technology disseminated through the Technology Institute of the Addis Ababa University (AAiT) facility, the Ethiopia 1550 Mushroom Program national mushroom spawn resources development centre, and the Ethiopia national mushroom technology training center.[91]

The tripartite AAiT, China Aid, and the Ethiopian Ministry of Agriculture mushroom project set up a standard mushroom spawn production, preservation, and technology training center, mainly for mother spawn production and teaching technology classes before being gradually expanded to include more nationwide training, advocating Ethiopia's mushroom production, productivity, and other benefits from the sector.[92]

The Ethiopian program constructed five mushroom project sites in its capital, Addis Ababa and its periphery to generate next-grade spawns, and Esayas Gebre-Yohannes (executive director of AAiT) stated the objects at the project's official opening ceremony at the AAiT, indicating the program's contributions to Ethiopia in advocating mushroom production and productivity, generating employment, technology transfer and research and development (R&D).[93]

---

[90] Tamfu, Arison, *Op.cit.*

[91] Xinhua, "Ethiopia teams up with China on mushroom project" dated 19 December 2019 in China Daily [downloaded on 19 December 2019], available at https://global.chinadaily.com.cn/a/201912/19/WS5dfaec4ea310cf3e3557f3e0.html.

[92] *Ibid.*

[93] *Ibid.*

Abdulsemed Abdo (adviser to the Ethiopian Minister of Agriculture) highlighted that consumer preferences and perceptions have not been well-researched despite mushrooms being a crucial food resource for the world and Ethiopia, and he advocated enhanced mushroom species cultivation with modern feeds for bigger yields through technological development:

> "Some of the specific challenges associated with mushroom production include the limited capacity and lack of role clarity of the different actors, the focus of the system on very few varieties, mismatch between supply and demand resulting in shortage and excess inventory, and quality issues due to inappropriate production, storage, and transport practices and facilities."[94]

The Chinese state has also indicated its public support. Economic and commercial minister counsellor at the People's Republic of China (PRC) Embassy in Ethiopia, Liu Yu, articulated her strong belief that more Ethiopian farmers would grow edible mushrooms under the Centre's supervision for generating bigger profits and better lifestyles.[95]

Osmosis of Knowledge

Exchanges in mushroom knowledge go both ways. While China has introduced Juncao technology to the rest of the world, foreign researchers and experts have also contributed to China's body of mushroom knowledge. Sri Lankan Professor Samantha C. Karunarathna (PhD, nicknamed "a mushroom born in Sri Lanka and grown in Yunnan") arrived in China to become a mushroom specialist of mushroom taxonomy, phylogeny, and domestication of mushrooms at Qujing Normal University.[96] In 2022, Karunarathna got onto the faculty at the Center for Yunnan Plateau Biological Resources Protection and Utilization, College of Biological Resource and Food Engineering at Qujing Normal University, instructing

---

[94] *Ibid.*

[95] *Ibid.*

[96] Long, Yun and Bi Weizi (with contributions from Kong Youqiong), *Op.cit.*

students while churning out specific informative mushroom-related scientific journal papers and publications:

> "It was estimated that there are approximately 900 edible mushrooms in Yunnan, and as far as I know, there is no [other] place on the planet with this high mushroom diversity . . . the colors and different shapes of mushrooms attracted him to gain more knowledge about "this interesting group. Compared with other countries, China offers a greater variety of opportunities. From this experience, I learned and am still learning how rewarding it can be to see scientific knowledge being implemented in practice and benefit or improve existing systems. I would also like to invest my energy and research to support the local farmers in Yunnan through mushroom cultivation and domestication. My students are highly diligent in their studies and often work hard day and night. [He hopes in the future, they will not merely view their studies as a means to secure jobs but genuinely have a passion for the study of wild mushrooms . . . Karunarathna firmly believes that passion is the best teacher because true happiness stems from the love of learning itself, enabling people to persevere in their scientific research, regardless of the difficulties they may face.] I strongly recommend anyone who visits Yunnan to try the [culinary] specialty of the province-Wild Mushroom Hot Pot. It is a unique experience you can only get in Yunnan."[97]

Karunarathna utilized a rationalized scientific approach to studying and utilizing wild mushrooms as a source of income for local peasants, tapping into China's significant investments in research and development (R&D), especially in protecting endangered species, rejuvenating environmentally worn lands for agriculture and benefiting from Yunnan's green energy, green food, and healthy living environment initiatives.[98] In fact, in general, mushroom farming is about the reuse of natural resources, as it also involves recycling organic matter for use in growing substrate in plastic bags or containers and then returned to the soil as fertilizer.[99] Agriculture, forestry, and food processing churn out big volumes of different types of organic waste and growing mushroom substrate

---

[97] *Ibid.*

[98] *Ibid.*

[99] Adusumalli, Harshini Priya, Nur Mohammad Ali Chisty, Rahman, Mahofuzur, Hossain, Shakawat Hossain and Pasupuleti, Mahesh Babu, *Op.cit.*

convert wastes and woods into commercially-viable materials as a form of sustainable agriculture and it can also prevent destructive forest fires by consuming overstocked small-diameter wood waste from wood manufacturing.[100]

Other than general research on mushrooms, there is also a specific environmental angle to the use of the much-touted Juncao technology. In 2017, the China-UN Peace and Development Trust Fund launched the Juncao Technology Project at the UN Headquarters in New York City with China contributing its experience and resources to the implementation of the 2030 Agenda for Sustainable Development, advocating the technology as a way to boost local incomes through affordable mushroom cultivation while arresting desertification.[101] For example, after China set up its inaugural overseas Juncao technology demonstration base in Papua New Guinea (PNG) in 2001, the Juncao technology has benefited 106 countries and regions by tackling poverty, soil erosion, and desertification.[102]

NGOs are also involved in Juncao cultivation in the CAR. Jean-Christian Ngakoutou-Patasse and the International Non-Governmental Organization (NGO) Centrolive that he leads endorsed and allied with the Juncao tech to train approximately 100 practitioners to generate revenue to get out of poverty: "We built the mushroom house and Juncao technicians brought us the mycelium that allowed us to develop this technology. We are a country coming out of a war, and today many young people are neglected. Through this training, we will help them to channel themselves towards a technology which allows them to do a lot more development so that they can support themselves. China is already giving us a lot of support, that's very good, we are counting on this cooperation to evolve together and help the people of Central African Republic."[103]

Such technologies are couched in state-owned media as a component of China's core leadership in constructing "a community with a shared future for mankind".[104] The 2014 Fiji Juncao project endorsed by President Xi and the leaders of the two countries is couched as the following global contribution by the state-owned media: "The wider application of Juncao

---

[100] *Ibid.*

[101] Global Times, *Op.cit.*

[102] Shen, Weiduo, *Op.cit.*

[103] Tamfu, Arison, *Op.cit.*

[104] *Ibid.*

technology in the world is seen as a vivid display of Xi's economic philosophy, as the seemly small but valuable herb has offered China's solutions to global challenges, nation's commitment to foreign aid as well as its concern for the greater good of humanity with an ambitious goal of "no one will be left behind" in the global development."[105]

President Xi indicated in his congratulatory letter celebrating the 20th Anniversary of Juncao Assistance and Sustainable Development Cooperation in 2021 and the First High-Level Conference of the Forum on Global Action for Shared Development in July 2023:

"I have long cared about the international cooperation of Juncao technology. [China is willing to work with relevant parties to continue to contribute China's wisdom and China's solutions to the implementation of the United Nations 2030 Agenda for Sustainable Development and make Juncao technology a "grass of happiness" that benefits people in developing countries.] Development is the eternal theme of human society, and shared development is an important path to building a better world. As the largest developing country, China has always placed its own development within the larger context of human development and created new opportunities for the development of the world through its own development. China will further increase its allocation of resources for global development cooperation, work with the international community to continuously deepen and substantiate the Global Development Initiative, and make new contributions to realizing the UN 2030 Sustainable Development Goals as scheduled and promoting the building of a community with a shared future for mankind."[106]

Overall, Chinese state media appeared to illuminate the positive results of the technology. Juncao appears to complement the overall growth of the Chinese mushroom industry. The average annual growth rate of the mushroom industry surged above 10% in China from 1992 to 2022 while adapting to the trends of awareness of food security and rural development through the use of sustainable practices using bio-innovation, technical digitalization, and marketing.[107] Growing mushrooms can provide sustainable and healthy dining experiences because

---

[105] Shen, Weiduo, *Op.cit.*

[106] *Ibid.*

[107] Adusumalli, Harshini Priya, Nur Mohammad Ali Chisty, Rahman, Mahofuzur, Hossain, Shakawat Hossain and Pasupuleti, Mahesh Babu, *Op.cit.*

of the decreasing number of wild mushrooms due to environmental pollution, natural resources, and costly human resources.[108]

In 2017, the China-UN Peace and Development Trust Fund launched the Juncao Technology Project at the UN Headquarters in New York and 2 years later, María Fernanda Espinosa Garcés (President of the 73rd Session of the UN General Assembly) said at the UN High-Level Meeting on Juncao Technology in 2019: "Through Juncao technology, China has a great story to tell – a story now shared with over 100 countries that have benefited from this innovation. The spark lit in Fujian Province has shown the potential of a single innovation – if nurtured and deployed wisely – to change lives and improve livelihoods across the world."[109]

The Rise of Another Emerging Economy, India, in the Industry

In the area of poverty alleviation, there appears to be much potential for all developing, fast-emerging economies. China is not the only player. In 2019, Fulrida Ekka from Siliguri in West Bengal wanted to get more income after her spouse passed away, and the part-time seasonal tea leaves-picking assignment was just not sufficient to support her household.[110] With the assistance of an Indian rural development institution, Live Life Happily, she began retailing 2–3 bags of her mushrooms daily, which are grown in large bags hanging from the ceiling, generating for her approximately US$92 (£73) a month.[111]

Cases like Fulrida Ekka and their domesticated small-scale mushroom cultivation businesses are also found in China. In the early 1990s, the Chinese Edible Fungi Association estimated that small-scale households produced 95% of all mushrooms.[112] China has also become a ranking mushroom consumer, even using them to substitute beef, and, because of stronger domestic consumption, Chinese mushroom exports shrank from exporting 80% of the overall mushroom cloud to less than 20% in the

---

[108] *Ibid.*

[109] Shen, Weiduo, *Op.cit.*

[110] Nandy, Anirban and Fulrida Ekka, *Op.cit.*

[111] *Ibid.*

[112] Adusumalli, Harshini Priya, Nur Mohammad Ali Chisty, Rahman, Mahofuzur, Hossain, Shakawat Hossain and Pasupuleti, Mahesh Babu, *Op.cit.*

early 2000s and below 5% by 2022 while allocating 80% of its production for its domestic consumers in the early 2000s.[113]

Such direct selling methods by small-scale cultivators can also be found in China. In China, quality/food safety guarantees and convenient super-mart availability facilitated major mushroom producers' market dominance, while agricultural food wholesale markets in the early 2010s started to expand mushroom acquisition outlets from large-scale integrated agricultural food supply wholesalers/merchants to rented supermarket counters to retail their goods.[114]

Rouf Hamza Boda, an Indian mushroom specialist who spent 20 years identifying 100 types of mushrooms across Jammu and Kashmir, argued: "India has all the required elements for becoming a super power in mushroom production. India has huge wild mushroom diversity. Lots of composting material, cheap labour and [it is] supported by diverse climatic conditions."[115] He has an equivalent counterpart in China. Sri Lankan Professor Samantha C. Karunarathna at Qujing Normal University (PhD, Mae Fah Luang University, Thailand) worked intimately with Kunming Institute of Botany, Chinese Academy of Sciences (KIB-CAS) to carry out fieldwork in Yunnanese mushroom diversity, and it led him to travel to Kunming before settling down at KIB.[116] In KIB, Karunarathna worked on a diverse spectrum of research projects with his colleagues to enhance the sustainability of agricultural systems and rural livelihoods, while Karunarathna's own personal mission was to identify new edible Yunnanese mushrooms for consumers.[117] Specialists like him are part of the ecology of stakeholders that makes the development of the Chinese mushroom industry possible.

Finally, outside very serious ventures and undertakings, mushroom cultivation can be fun too. Some argue that mushroom cultivation may even become a lifestyle choice as a "feasible and pleasant hobby for both rural and peri-urban residents" as it does not need a sizable outlay of capital, is bereft of intense maintenance, manageable part-time, and can

---

[113] *Ibid.*

[114] *Ibid.*

[115] Nandy, Anirban and Fulrida Ekka, *Op.cit.*

[116] Long, Yun and Bi Weizi (with contributions from Kong Youqiong), *Op.cit.*

[117] *Ibid.*

be expanded from hobby to business according to capital and human labor availability.[118]

[For more comparative angles with other East Asian cultivators of high-tech mushroom crops, please refer to Arthur Pedida's chapter on the Singapore case study of high-tech mushroom cultivation in this volume.]

---

[118] Adusumalli, Harshini Priya, Nur Mohammad Ali Chisty, Rahman, Mahofuzur, Hossain, Shakawat Hossain and Pasupuleti, Mahesh Babu, *Op.cit.*

Conclusion

Tai Wei Lim

Abstract

This chapter reiterates some of the major points indicated in the country chapters and points to certain trends or recommendations for readers to ponder upon.

The case studies in this volume started with the detailed analyses of Pedida on Singapore's mushroom farming industry. As the population continues to rise, urban farmers in Singapore present some challenges to support the country as a self-sufficient nation. Urban farming gained importance as a sustainable solution to a confined metropolis like Singapore, with an increasing public interest in locally sourced food. Local companies are utilizing various farming techniques using vertical farming, hydroponics, and aquaponics, and they've created a multi-dimensional, innovative farming solution. Local farmers, growers, and the rest of the communities are actively involved in cultivating a wide range of crops within the city, contributing to meet their dietary needs, food security, and environmental sustainability. Pedida makes a strong case for high-tech farming in a densely populated environment.

In a similarly high-tech environment but with strong references to historical traditions and customs, Godo's chapter begins with a brief overview of industrial revolutions and how they have transformed Japan's industries and agriculture. This useful backgrounder then proceeds to explain how this industrial process transformed the agricultural industry from one of comprehensive lifestyle materiel provision to more focused food materials. Kenji Maeda's herb farming is featured as one example of the balance between these two objectives and goals, something made possible with the use of net-based awareness/knowledge dissemination solutions to ICT (Information and Communications Technology) tutorials.

In his introduction to scientific farming in Japan, his sub-sectional case study on the Maeda Farm discusses a new type of "smart agriculture". It studies how farmers like Maeda created a new business model based on the growing new communication technology.

This was followed by a chapter that first provided the historical background behind Japan's agricultural land reforms and detailed how land reforms may benefit from Industry 4.0. Godo argued that, unlike mechanization before Industry 4.0, artificial intelligence (AI), combined with information technology (IT), enables flexible thinking that can make more flexible and dynamic use of agricultural land possible.

As with the previous chapters, Godo provides background information on the geographical specs of Japan's marine spaces. He then proceeds to detail the modern features of Japan's fisheries and fishing industries. The coverage is detailed, even touching on work ethics, culture, and legislative acts (including examples of deregulations). This chapter advocates how updated legislations have made it possible for IoT and AI, as well as other Industry 4.0 technologies can make it possible for efficient record-keeping, big data generation, and maintenance of databases possible for all functions, from monitoring water condition to protection of endangered species.

Finally, Lim's chapters on China briefly survey the history of technological use in Chinese agriculture and food sectors. It adopts area studies and political economic angles to examine this topical matter with a focus on policy analysis related to the utilization of Industry 4.0 technologies. China appears to be a combination of both the desire for the utilization of farming tech in densely-populated areas (similar to Singapore but on a larger scale and the world's second largest population), as well as having some references to the specific agricultural landscapes/histories/cultures of the country (much like Japan).

With these three case studies pegged at different scales but with the same intense desire for engaging Industry 4.0 technologies in the agricultural and farming industries, brief insights into East Asian agricultural/farming practices are provided for the reader. It does not pretend to be comprehensive but provides some overviews and contours of the agricultural/farming industry to the readers through practical case studies.

Index

9 789819 830190